THE PSYCHIC NATURE OF PHYSIOLOGY

THE PSYCHIC NATURE OF PHYSIOLOGY

John M. Dorsey, M.D.

Published by
Center for Health Education
4421 Woodward Avenue
Detroit, Michigan 48201

Printed by Edwards Brothers, Inc.

*Not every scientist is supported like Kepler
by the feeling that in discovering the ways of nature
he is "thinking God's thoughts after Him"*

John Baillie
Natural Science and the Spiritual Life, 1951

CONTENTS

FOREWORD

The office of the University Professor may be readily conceivable as that of observing and understanding the universal in the particular, the man in the mind, the real unity of mentality underlying its apparent plurality. Over the years Dr. Dorsey has met this scholarly obligation head-on, shouldering it firmly and perseveringly.

His first book addressed to this organic conception, *Living Consciously: The Science of Self*, appeared in 1959. I co-authored this volume. It already foreshadowed clearly this present production. Each of his subsequent publications reflects his consistent growth in the natural direction of honoring his self-understanding as the only basis for all of his knowledgeability. Every title tells the story of the development of his self-insight, beginning with his living monument, *Illness or Allness* (1965), and followed by each of his *Psychology* texts specifically describing this psychogenesis: *Language* (1971), *Emotion* (1971), *Political Science* (1973), *Ethics* (1974).

The author's first book, *Foundations of Human Nature: The Story of the Person* (1935), was a cultivation of his notable appreciation for individuality. His editing of *The Jefferson-Dunglison Letters* (1960) repeatedly scored the helpfulness of conscious individualism. And his most recent publication, *An American Psychiatrist in Vienna, 1935–1937, and His Sigmund Freud* (1976), provides a fine account of how he grew his consciousness for his self-development.

Claude Bernard emphasized the sameness in experiments with living materials and those concerned with liquids, solutions, and instruments in the physical chemical laboratory, but he was not ready to acknowledge personal scientific authority. He knew that certain mistakes are inherent to the period, and great truths surface in ordered sequence. John Dorsey was ready to advance his insightful observation of acknowledgeable self-conscious responsibility, appreciating all of his creativity as his self-conscious personal living. To a physiologist this mental development reveals new horizons, observing that his own responsibility as an individual subsumes his professional and private life. It is a great advance to recognize personal scientific authority. Indeed, professional and private living are aspects of the Self in action. The Self is a proper and the only possible study of physiology. Like all else, physiology is of, by, and about the person as an individual. Trying to conduct one's physiology as if this were not true accounts for the genesis of pathological physiology.

As a physiologist devoting my life to research upon appreciation for its functioning, I hail this latest publication from the life-describing pen of my esteemed colleague.

Walter H. Seegers, Chairman, Department of
  Physiology
    William D. Traitel Professor of Hematology and
      Director
        Thrombosis Specialized Center of Research,
        Wayne State University

ACKNOWLEDGMENTS

I wish to acknowledge my indebtedness to those who have helped to make this publication possible. First of all, I thank each of my Trustees of the Center for Health Education.

Dr. Dwight C. Ensign, Chairman
Mr. H. Walter Bando
Mrs. William D. Crim
The Right Reverend Robert L. DeWitt
Mrs. Carl B. Grawn
Mr. Robert F. Grindley
Miss Emilie G. Sargent, R.N.
Walter H. Seegers, Ph.D., M.D. (Hon.).

I duly acknowledge the special kindness of Barbara C. Woodward for her reading my manuscript and making valuable observations about it.

I am glad to include the artistry of William E. Loechel of Wayne State University who depicts insightfully his conceptions of Growth and Experience.

I thank my secretary, Nellie Richardson, for her great patience and skill.

I am ever most grateful for the finest attention and comments of my wife, Mary Louise Carson Dorsey.

The symbolic emblem of the Physiology Department

of Wayne State University on the front jacket and cover of this book needs no explanation. Its four qualities of mind: *good nature, integrity, industry,* and *science,* in this order of their importance, were written by Thomas Jefferson to Benjamin Rush, the "father of American psychiatry."

I feel convinced there will come a day when physiologists, poets, and philosophers will all speak the same language and understand one another.

Claude Bernard

INTRODUCTION

Bernard, great physiologist that he was, here obviously understood the purely psychical basis of all language. In this prophecy he came very close to recognizing how each scientist is already talking a language similar to every other one, namely, the language verbalizing the meaningful experiences of his own life.

First of all, just what do I mean by "psychology"? My psychology is my *growing* of the study of my mental functioning, the study of how I use my mind, the study of the meaning of all of my meaning. My "study," itself, is a psychic function. Thus, all of my knowledge is and must be my self-knowledge, and it is healthful to fully appreciate that whole-making truth. However, I may not have developed my mind with that degree of understanding, so that I may tend to bypass the basic fact of my own life in creating my meaning for each of my studies, even for my study of psychology itself.

Whether I use my mind for study or not, it goes on functioning just the same. My appreciation for this truth may be strong or weak. To illustrate, I may fully acknowledge that all of my meaning must be psychic in nature but also go on to posit 1) pure psychology, by which I mean the use of my mind for studying my whole mind, 2) applied psychology, by which I mean the use of my mind for studying a specific area of my

1

mind, 3) unstudied psychic activity, by which I may mean merely living my stream of consciousness with or without any purposeful observation of its flow of meaning, or 4) unconscious psychic activity, by which I may mean my mental functioning while I am asleep in varying degrees of depth.

And just what do I mean by "physiology"? My physiology is a specific instance of my using my mind to study my mental functioning with specific reference to my bodily activity, as such only. However, the biological process of my whole organism includes every meaning that my mind can grow. Nevertheless, for purpose of definition, I can arbitrarily concentrate my mental growth in one or another area of my self-interest, such as physiology, thereby creating the illusion that I can isolate it from its erstwhile organic continuum. However, I find genius ever identifiable with abundance of whole-self consciousness. Study of my physiological functioning, for example, may be pursued *as if* completely separate from the several meaningful areas of my other organismic conduct, such as my sociology, zoology, philology, ecology, theology, etc.

I propose the conception of "objectivity" as the most consequential fiction ("as ifness") in the history of thought, undeniably necessary and useful though it may seem to be.

I must fear using the term *solipsism* freely until I recognize my own wholeness sufficiently to begin to understand that my life must be, solipsistically, all and only my own life. Then I find this same word helpfully descriptive for my intact individuality.

I wish to specify exactly what I mean by the term *solipsism*, since I find many meanings ascribed to it. To me, solipsism signifies merely: whatever is, is entirely and only itself, by virtue of how it has come to be.

2

The word itself explains itself: *solus,* means alone or one; *ipse,* means self.

There can be no separation in a life, only wholeness-unity. "Separation" is a word that bridges over acknowledged self and unacknowledged self. "What is the meaning of my life?" is the question of questions for everyone. As Einstein said once of the universe, I say of my own existence, namely, the most incomprehensible thing about my life is the fact that I am finding it comprehensible. This self-understanding is resulting from my growing sufficient self-acknowledgment to be able to observe that my life, alone and as such, is my all, my indivisible oneness or wholeness.

To know what my life means I must first acknowledge that the existence of that meaning is created in and of my mind, then study its power and discover the precious benefit in looking it over as well as the costly risk in overlooking it. *Psychical* reality is the only possible subject of every scientist. My mind is my only possible realm of my every so-called datum.

The following fact must be fully honored: Attempting to translate mentality into so-called non-mental body introduces the problem of trying to understand internality through externality (that is, subjectivity through objectivity), whereas each of these dual terms applies strictly and only to a set of meanings *purposely* intended as if to exclude its other one. Analogously, the nomenclature of every science or special discipline is constructed *purposely* as if to exclude the meaning of every other science or interest.

My mind is the living of my meaning of all of my experience. Whatever my mind lives and may mistake for an external event is only itself specifically modified for such special "as if" *purpose.* Since my acknowledgeable mind can bear only so much responsibility for being itself at each stage of its development, every kind and degree of denial of being itself (conscious mind's repudiation of its own experience) thus serves the growing mind's purpose.

Each so-called illness, every symptom, each so-called accident is also *purposive.* Whether it is named "organic" or "functional," my every degree of so-called ill-health must be understood as my only present way of helping myself to stay alive, hence as desirable and honorable. Otherwise each symptom serves the *purpose* of concealing my unconscious living, of preventing my unconscious mental power from becoming my conscious mental power. To illustrate, I have detailed elsewhere my theory of cancer as being my mind's purposeful effort to prevent itself from becoming overwhelmed with responsibility for acknowledging its alienated unconscious power (concealed in any and all so-called "otherness") as being its own.[1] A fully "overwhelmed" mind is disorganized, its acknowledgeable orientation of personal self-identity becoming suspended. Hence, it is feared as death itself.

It has required sufficient self evidence for me to be able to convince myself of the certainty that any and all disease can serve the *purpose* of maintaining my mind's *status quo,* specifically by preserving the current balance of my conscious mind and my unconscious mind. For example, if I have rejected (*as if* alienated, made unconscious) such natural wishes as being loved by somebody else, being dependent upon somebody else, being angry at somebody else, being grateful to somebody else, and so on, then by becoming sick enough I can justify my resorting to all such natural infantile behavior.

My own esteemed medical teachers seemed to be devoted to the nineteenth century materialism, the product of consciously inhibited idealism. Henry Margenau, Yale's Eugene Higgins professor of physics and natural philosophy asserts: "Materialism was a respectable philosophic view at the end of the nineteenth century; it has now become an anachronism."[2]

Wondrous wise Demosthenes declared: What we wish, that we believe. However, what I am including in what I believe (as in all else that I live) can be defined only in terms of my historical development. Only after I learned how to investigate intensively and extensively whatever I could not acknowledge to be my own mind, could I make my habitually disowned inductive method of research (namely, observation itself) the subject of my investigation, thus discovering it to be all and only my *self* observation.

The need for responsible self-acknowledgment of my anatomical and physiological creations was mostly obscured by the immediate necessity that I grow a large vocabulary for merely describing these thick coming, numberless developments. I seemed to lose myself by making thousands of new words and meanings my own, without appreciating each one as being all and only about my own rapidly growing mind. My study of the law governing the growth of my mind finds its most useful clarification in the understanding of language I term idiolect.[3]

Applying my historical perspective of growth to my own life has been the sole source for cultivating my self understanding revealing to me the psychic continuity of my inventions and ideas of self-helpfulness. That I can only *grow* as myself whatever I sense, perceive, feel or know, is most essential insight for my appreciating the wholeness of my individuality. My science of psychology must be based on my knowledge about how I have grown, and am growing, my own mind.

I can secure information about the fundamental traits of my own mind only, by making *consciously present* my reconstruction of what I conceive to be its eventful history. This task of self-understanding is well worth my most strenuous effort, for all of my mental health trouble is traceable to my assuming that what does exist only within me can either influence, or be influenced by, another one that must exist entirely and only within *his* (*her*) own selfness. *My* anthroposcience is the true domain of my every systematic investigation. The only fact I can ever establish must be the psychical one that I observe (conceive, make).

Meaning is psychic functioning, but lacking the understanding that I must be totally responsible for minding my own otherness-meaning, I am apt to think of myself exclusively as a body separate from innumerable other bodies that are not my own. Conventional confusion!

From this increasing ignoration for the necessity that I must mind (live mentally) my own *self-grown* otherness-meaning, I can derive the notion that meaning can be non-mental. From then on it is an easy step to identify myself as a physical (non-mental) body. Soon I become adept in contriving names for the innumerable subdivisions I make of my own mind, attributing my meaning of impersonal embodiment to all of it, a body of people, of water, land or any so-called aggregate that seems to be able to lump individuality with individuality. Only my own systematically conscious self-analysis undertaken with my Professor Sigmund Freud provided me with the experience of self consciousness enabling me to construct my consciously psychic psychogenesis.

My much-worked illusion of "relationship" asserts plurality and thus gainsays the necessary innerness, wholeness, and unity of individuality. Presumably mind is predicated upon life so that the implied unity in the word "biopsychic" seems scientifically admissible. I take for granted that habit of mind prevents my use of my mind for growing only evident personal identification in my own experience. Thus, I may posit that the meaning of biology enables the meaning of psychicality, on the assumption that life comes first. However, all such reasoning is purely mental, so that all that can be meant by my biology is based upon my psychicality only. *All of science is psychical theory.* My ideal is my only real.

With this all-important understanding, for alluding to implied unity of my psychical and any other life interest, I use the term *psychical physiology* or *psychobiology* rather than *physiological psychology* or *biopsychology.* In this sense, the concept "psycho-social" honors necessary intraorganismic experience whereas "social psychology" may not seem to do so.

My ardent naturalist colleague demands an "objective" connection called "relationship" between the phenomena he studies, which so-called subjective entities seem to him to lack because (my) he has not been in the habit of heeding his creation of his united wholeness. Goethe observes,

> Every fact, itself is the really interesting object. Whoever explains it, or connects it with other events, usually only amuses himself or makes sport of us, as, for instance, the naturalist or historian. But a single action or event is interesting, not because it is explainable, but because it is true. [*Unterhaltungen deutscher Ausgewanderter*]

Claiming to be an objective scientist only, I can exercise my omnipotence of thought by assuming myself to be capable of doing research on my external world. I choose, however, to heed my subjectivity as the only source of any proposed objectivity, because of the healthfulness in recognizing the whole of *my* world in my individuality, and because of the hindered healthfulness in localizing the whole of my individuality merely in an inconceivably larger totality that seems to exclude or belittle my identity.

It is the whole of my being, to which I ascribe the law of my being, that must compel my interest and

7

devotion. My recognizing my own individual unity in my "as if" objectivity, as well as subjectivity, is indispensable to my honoring such wholeness of my wholeness. This perspective hardly affords a "possible object of study" for my standard naturalist who depends upon his pluralities, fractions, personifications, comparisons, relationships, and so on, for his psychical productions.

I note that the whole view of this scientific fact, that *meaning* is always a psychical construct, discloses my Mind-Body problem to be a phantom problem, created by my overlooking the observation that all that I can *mean* by body must be mental.

With this insight there began to dawn upon my conscious self a feeling that I can grow my mind as an aesthetic observer of self-acknowledgeable significance. As I continued to develop in this direction of conscious mind responsibility, this feeling (of being all of my experience) grew into a conviction. Coincidentally I understood why my esteemed fellowman, as a rule, in his reporting his life history touches upon his conscious mind-life only incidentally: that would seem merely inconsistent with defending his posture of objectivity maintained by exalting his ideas and feelings of so-called unselfishness.

However, in spite of my systematic scope and my thoroughness in the field of *as if* objectivity, there remains a distinctly urgent need and ubiquitous place for explicit and unremitting emphasis on this lifesaving truth: all of my "objectivity" must have its whole origin, course, and termination in my *subjectivity*. In fact, then, all of my "objectivity" must be my unrecognized (unconscious) subjectivity. "Lifesaving" because maintenance of the self-belittling way of life (implicit in my systematically practicing self-disregarding "objectivity") requires symptom-formation warning me of its life endangering consequence.

I trace high mortality in my profession directly to this distressing necessity. The medical student, as well as physician, suffers from it in his *morbus medicorum.* Frequently, to avoid awareness for this iatrogenic etiology, my colleague seems to jeopardize his life by indulging *as if* impersonal meanings for his chosen professional practice.

It is therefore with studied self-confidence that I compose this mighty little book as a record of my arduously using my understanding of the illusional nature of my face-saving "objectivity" in order to attain to my understanding of the factuality of my lifesaving *subjectivity.* Oh, for a life of conscious subjectivity that includes my unconscious objectivity! All of my habit of living objectivity otherwise is restful but represents a degree of sleep.

My own pre-insightful scientific examination of my mind's development reveals persistent effort to learn of it by studying whatever I could not acknowledge as being my mind, for example, *as if* otherness or objectivity. Thus I repeat in my ontogeny the developmental route of unconscious depersonalization taken by my phylogenetic forebears.

My gentle reader objects, "By stressing the idea that your physiology is psychical, for example, do you mean that when I try to look at myself in the mirror that I must include not only the figure of the person I see but also all that surrounds that figure, such as the chair or table or window or whatever beside me and behind me?"

When I do create my seeing of my mirrored figure I also see my mirrored chair or table or window or whatever structure is beside that figure, and it is also that latter seeing of mine which is justly acknowledgeable as entirely my own inarticulate, seamless, organic mentality.

"I understand that distinction. My mirrored outline appears to present me as being separate from my environment, as a body among bodies, whereas my true image must be a creation of my mind which my mirror cannot reflect. My mirrored form seems to support my illusion of objectivity, does it not? Is your empiricism purely your own developing mentality, your doctrine of growing your self experience?"

Yes. By using the evidence of my visual sense, or any other sense, without acknowledging it as evidence of my own mind, I create my illusion of objectivity. My mirror cannot reflect the shape of my mind beyond that fashioned in the lineament of my so-called features or build. But even the mentality of my body is not discoverable in my mirrored contour, unless I begin and sustain that visual observation with the understanding that all of it must be mental from start to finish.

"May I test further your idea of psychicality-physiology unity, or identity, or wholeness? Do you mean that all that I call argument or conflict of ideas is really nothing but verbalized organic tension analogous to my tensing my agonist and antagonist muscles? Do you mean that my mind's meaning, for example, my thinking on externality or psychology or philosophy, is indispensable for my organic integration in the same sense that my bones and tendons and muscles are? Is every organ of your body constantly expressing the way you conduct your life either consciously or unconsciously? Does the condition of your lung or liver or heart depend upon the way you teach yourself to manage your wishes and fears, and *vice versa*? Does your attitude towards your sex life have much to do with your disciplining yourself in self-control"?

To each such question of mine, on every count, I must say "Yes." First, because every question (as every other mental event) is itself life affirming in nature.

10

Second, because each of these questions consists of meaning that is consonant with the overall truth of self-continence, of absolute individuality.

"You must mean that my saying 'No' to *any* of my experience (my mentality) is to assert: What is, is not: or What is not, is. For example, if I say that my body sensations are not mental, then I am thereby denying the existence of the body nucleus of my mind."

Yes, as Sigmund Freud stated, simply to observe how a body need, elimination for example, can become an overriding personal concern convincingly demonstrates body-ego priority. Unpleasing emotionality and painful body-sensation can readily be observable as congeneric. To illustrate, my experiencing any pleasing or unpleasing social happening can correspondingly alleviate or aggravate any painful sensation I may have been feeling. Certainly my enjoyable socializing and pleasing body sensation can be observed to be congeneric. Thus, my gratifying social event can facilitate my elimination, appetite, exercise, or other sensory pleasure.

"My common sense tells me that my sight, hearing, touching, and so on, enable me to sense my world, but do you mean (or observe to yourself) that your senses only enable you to discover your own organic identity or unity? Is all innerness in itself only? When I think I am seeing my fellowman, is that decision a visual hallucination based on an optical illusion?

"Do you think that my beliefs or prejudices or decisions or perceptions are organically influential in my metabolism or appetite or elimination? Can you observe in your own experience how your love life determines the state of your health or how your likes and dislikes enter into the determination of how long you are going to live? Do you have to appreciate your wholeness-nature in order to cultivate a biologically adequate understanding of the meaning of your life?"

When I ask myself each of these questions the answer is uniformly: Yes, certainly. My organic nature is that of a whole continuum, not that of connected parts. In fact all so-called differentiation is nothing but inhibited individuation. Throughout all of my extent only my universal wholeness obtains, whether consciously or unconsciously. Liberating my mind from its inhibited freedom consists of my developing my unconscious wholeness into my conscious wholeness. The only possible solid basis for establishing physiology is the physiologist's own mind.

I describe a welcome economic turn of my mental development in which my imagination enters to "master matter by means of the idea" (Keats), so that the nature I feel in my life can be gladly and sensibly possessed as truly my own. It is invigorating to be able to feel each of my perceptive, intuitive, cognitive, conative, and affective processes as a sensation of my mind. I long for further progress in this mind consciousness, for it does lighten the load of self responsibility to be able to know precisely where the burden lies. Whatever the subject, I can think and feel it with the peace of right spirit by comprehending it as my own necessity for consecutive growth.

I expose how I grew the insight about what could make me identify my self despondently as a sufferer: by heeding how the diremptive power of my mind functioned without my volitional control, making me appear to myself as if I could be torn asunder by my own acknowledged and repudiated powers. Therefore I could not interpret my suffering as my growing understanding of the deeps of the soul, as well as my developing the virtue of mind most essential for fulfilling my nature. Insightful endurance of the specific helpfulness of all pain and sorrow draws my appreciation to my need to study and obey carefully the law of my biologically adequate way of life.

It seems safe to say the content of this slender volume may be tolerably understood by a beginning student of any of the health disciplines.

What's done we partly may compute
But know not what's resisted.

Robert Burns

PERSPECTIVE

For the purpose of highlighting the revealing *necessities* accounting for the seemingly slow growth of insight discovering the psychic nature of physiology, I begin with a perspicuous medical study written at the beginning of this Century. In his glowing book, *The Force of Mind; or The Mental Factor in Medicine*[1] the spirited British physician, Alfred T. Schofield, described the tremendous resistance the mind ordinarily exerts against considering the mental nature of health and its vicissitudes. What he stated then has much the same relevance now:

> Though the leaders in the profession have recognized the importance of the mental factor in medicine in all ages, it is generally ignored today.

As might well be expected, the occurrence of this disregard can be fully justified. Furthermore, it is most enlightening to study the truth of it, for in no other way can I renounce an unreasonable attitude of fault-finding about it.

Dr. Schofield quotes from Dr. S. Weir Mitchell's *The Physician*:

> There are among us those who haply please
> To think our business is to treat disease,
> And all unknowingly lack this lesson still,
> 'Tis not the body, but the man is ill.

17

Schofield also quotes Dr. T. Laycock, teacher of Hughlings Jackson:

> The most eminent and successful physicians have all been psychologists; for a knowledge of a practical science of mind is fundamentally necessary to the practice of medicine.

In his chapter on "Psycho-Physiology," Schofield records, "The whole body is the organ of the mind." "We live consciously and exist unconsciously," "Where 'conscious' ends unconscious begins," and "The mental factor consists *generally* in carrying on the functions of life; and *specially* in physically expressing mental states." His chapter, "The Vis Medicatrix Naturae," is as practical today as when it was composed, offering authentic testimony of the body's therapeutics, beginning with the words of self-heal of Hippocrates, "Nature is the physician of disease."

I write of the life of insight as being the ideal life of life. Many an ancient has recorded the same. Why do I? On two counts: first, I know of nothing more critically important; and second, I know of nothing more critically resisted. Furthermore, I know of no greater resistance to insightfulness than in my study of physiology. By *insight* I mean specifically: appreciation for the psychic nature of all of my meaningful experience. Insight, responsible self-observation, mind-awareness, self-understanding, conscious self-knowledge, acknowledgeable individuality, recognizable self-growth, continent life appreciation,—all are cognate expressions.

Sigmund Freud built his whole theory of psychoanalysis on his perception of resistance offered by his patient to considering areas of his mind he had associated with pain or unpleasure in some form and, hence, had chosen to exclude from his preferred self-conception. The distressing feelings making up my resistances (like fear, guilt, shame, jealousy, hate) serve to help me to avoid the activation of my mind that I associate with the source of the distress, quite as tenderness helps me to keep from aggravating any other kind of wound. Thus resistance demonstrates itself as maintaining a *dissociation* of my mind into 1) that which can function freely and 2) that which is inhibited in function (technically termed "repressed"). Protean mental symptoms are derived from thus interfered-with mental functioning, each symptom being a compromise between the original function itself and the functioning of the resistance opposing it.

Clearly, my symptomatic mental functioning specifically obscures my appreciation for the truly indivisible wholeness of my organic being. Clearly, it is my self-consciousness that I must work with in order to try to reclaim as my own the self-experiences I have formerly tried to disown. Clearly, my aversion to the very meaning of *psychic* becomes understandable as my defense against overwhelming my acknowledgeable self-identity with uncontrollable mental power. I must suffer from conscious self-minification from such power as long as I must continue to consider it as being alien to me. It is my self-consciousness only that can illuminate my mind to observe what William James called "a wider self through which saving experiences come."

In the following self-observations I attempt to explicate the role of psychic resistance in maintaining life itself. I try to clarify how 1) the amount of helpfulness of my psychic resistance is in direct proportion to the degree of its force, 2) the extent of symptom formation varies with the strength of the resistance, and 3) the psychic need for renunciation of the resistance necessarily corresponds with the amount of resistance present.

1) My resistance to any new mental development can be lifesaving because, without it, my mind may be overwhelmed with excitation so that it cannot maintain

its present level of conscious self-identity. It is my appreciation for my self as being myself that enables me to take any care of my life at all, so that I must avoid my losing any sensing of personal identity if I possibly can.

2) The greater my resistance to sensing any of my psychic experience with kind composure (that is, with assenting conscious self-continence) the more I must depend upon my symptom formation deriving from my need to live that peace-disturbing experience as foreign to my own acknowledgeable mentality.

3) The more I must depend upon my symptom formation for maintaining my repudiation of some of my own mentality, the greater the danger to my life. All of my repressed mentality continues to function, faithfully warning me with signs and symptoms of the peril in my disregarding the intact organicity of my wholeness-being. My lightly spoken, "I would rather die than think" this or that, may contain truth that I resist recognizing.

The strength of my resistance now becomes understandable as the strength of my need to disown some of my difficult experience. The symptom-forming alternative is a constant monitor reporting my need to cultivate an additional amount of acknowledgeable self-identity that can make room for whatever I have necessarily tried to exclude.

My lack of the feeling of completeness, of self sufficiency, is vague but it is safely interpretable as signifying an accumulation of my own experience that I live *as if* alien to my nature. This normal feeling of biological inadequacy is immediately dispelled by my feeling and realizing that *all* of my living is my own. As long as my understanding of ownership seems bereft of its psychic nature, I must suffer the wound I create by necessarily exalting my magic illusion: What is, is not.

My quality education is the conscious increase of my

20

possession of my mind for its own sake. Acknowledgeable subjective wealth provides riches that are true to life. I can be true to my love, to my fellowman, to my God, only as I can be true to my self. Whatever experience has been mine remains as a way to this conscious reality. My body is not an instrument of my mind, or soul. It *is* mind, it *is* soul.

Again, I turn to the beginning of the Century to illustrate the seemingly slow rate of growth of (my) self-insight. In one of his books of wisdom, Henry Wood describes the "supreme lesson" of life to be the development "of self-conscious individuality including a developed recognition of oneness with its source."[2] I hold his following ideals firmly in view, thereby actualizing each one. I also understand as completely natural my need to resist seeing my own identity in such unorthodox psychology.

> We patch the body from the outside, hoping thereby to make it mentally tenable a little longer. *But it is built from within.* From the inner to the outer is the universal established order.

And,

> But a small part of mentality is upon the plane of consciousness.

And,

> What one thinks most about, he either becomes or grows like, and it is the tendency and function of the physical organism to mirror it forth.

> No man has ever seen his friend, or even himself. It is the unseen which is the real and substantial.

> The ego has surrounded itself with a thought environment of discordant vibration with the established order.

> No drug by mysterious magic can remove penalty.

21

All visible phenomena are symptoms and effects.

ᴗause of its being common, abnormity is rated as normal, oɪ "natural."

The result of a six month's trial of pure scientific mental gymnastics will be bcth a surprise and a delight.

Whatever is held in the individual consciousness is in the deepest sense present and real.

All great advances in their earlier aspects have been rated as irrational innovations.

Man's mind is a busy factory where conditions are positively manufactured. He weaves their quality, consciously or unconsciously, into every nerve, muscle, and tissue of his body.

Daily psycho-gymnastics is needed, and is as important as physical exercise.

Negatives are not entities, but only deficiencies.

Man's idea of God is the very cornerstone, not only of his wholeness and happiness, but of his very being.

The original source of pain is always mental.

My physiology is not opposed to my mentality but rather is integral to it. It is able-bodied identity to be studied, trained, controlled, and made helpful in my subjective wholeness, a mental condition rather than an objective locality. This life orientation exalted and held steadily in mind invigorates my wholeness, continuing its ideality into the subsoil of my visceral subjectivity. Thus the wording of my idea does not stop at poetic imagery but unites it with the scientific discovery that all identity must be self-identity. It is vivifying to see the modern scientist finding his long sought self-sameness at last in mathematics, or in another less obscure form of metaphysics. The scientific idealist can make manifest the psychic in every discipline or interest, the identity in thought and thew, the divine oneness in pain and pleasure, harm and help, or bad and good.

The day of days, the great day of the feast of life, is that in which the inward eye opens to the Unity in things, to the omnipresence of law;— sees that what is must be, and ought to be, or is the best.

Ralph Waldo Emerson

GROWTH

SOME HISTORICAL BEGINNINGS OF CONSCIOUS PSYCHOPHYSIOLOGY

Elsewhere I have recorded "My Historic Evidence of my Conscious Self-development."[1] Nevertheless I include here mention of a very few life-conscious individuals, each of whom discovered and recorded the unique enjoyment of the power of living *consciously*. I find it vitalizing experience to try to trace to its historic origin the biological appreciation of individual mind as the origin course and termination of all (consciously or unconsciously) meaningful experience. To know *that I know* my Anatomy or Physiology or Biochemistry is to feel my unity in the meanings of each discipline. The physiology of my psychology is the psychology of my physiology. The flesh of my spirit is the spirit of my flesh. My unity of my innerness is all that I *can* experience in any way possible, consciously or unconsciously.

This revered insight enters into the understanding of the earliest acknowledged idealist of the East as of the West.[2] The founder of every great religion sensed (minded) it. The ancient Sanskrit word, *ahankara*, denoted I-maker, the principle generating self-consciousness. Hippocrates (460-377 B.C.) according to Plato (428-347 B.C.) said that nature, even of the body, can only be understood as a whole. Anaxagoras, fifth century, B.C.,

27

affirmed mind (*nous*) to be the foundation of existence: "Things in one universe are not divided from each other."

My mind is my only living of any of the meaning of its activity. *Energy* is the Greek term precisely naming the operation of activity. It is only my sensing my consciousness that can report my sense of personal identity. Although I cannot be aware of all-of-myself except in my every conscious activity, I can be aware of the truth that my total wholeness does exist.

My act of consciousness is my mental modification that enlightens my mind so that it can be sensed for what it is, quickens my mind so that it can be felt for what it is, and frees my mind so that it can act for what it is, namely, the whole meaning of me. I discipline my mind by exercising its activity in the direction of continuous attention to its wholeness performance.

Aristotle (384–322 B.C.) recorded, "The intellect is perfected, not by knowledge but by activity." By its own activity of creating the *knowing*, to be sure. Said Thomas Aquinas, "The intellect commences in operation, and in operation it ends." Aristotle held that mind and body are one. Conscious self-responsibility and conscious authority constituted Socratic wisdom. Honored self-integrity is the open secret of the wholeness of life that can be revealed only by flash after flash of acknowledgeable self-possession.

Plotinus (A.D. 205–270) deserves the honor of being generally considered one of the greatest of the mystics. To me "mystic" meaning is most rarely understood as the individual's unswerving devotion to the wholeness and allness of his individuality. Plotinus's *Enneads* are masterpieces of conscious single-mindedness.

John Duns Scotus in the ninth century indicated awareness of the significant role of fictions, called *ficta*, in otherwise factual literature of the Middle Ages. I consider "objectivity" to be of this description, a fiction.

Here I wish (am driven) merely to allude, also all too briefly, to a few of my less ancient proponents of the effort to reveal this whole-making truth: the heeded organic integrity and invariable innerness of man. From their diaristic (conscious or unconscious) recording I continue to help myself.[3]

With due respect to every fellow scientist, whether subjectively disciplined consciously or not, I begin with Roger Bacon (1214-1294) who pointed *in: Original research is distinct from the acceptance of authority.* After all, there must be as many scientific methods as there are scientists giving life to their creations.

William of Occam (c. 1300-1349) grew great understanding of the fictional nature of many so-called practical ideas, nonexistent theoretically.

Francis Bacon (1561-1626) records in his *Novum Organum*, "The true and lawful goal of the sciences is none other than this: that human life be endowed with new discoveries and powers." My every science is a helpful, unique, godlike phantasy added to every product of my mental creativity that preceded it.

Thomas Hobbes (1588-1679) grew penetrating insight in his theory of Fictions, thus exposing *as if* illusions and making a way for further understanding of the wholeness of individuality.

Robert Burton (1578-1640), clergyman, wrote *The Anatomy of Melancholy.* Dr. Samuel Johnson said it "was the only book that ever took him out of bed two hours sooner than he wished to rise." Burton helpfully relates depression with irresponsible conduct of life.

René Descartes (1596-1650) distinctified mind and body, theorizing their interaction, however, in the pineal gland.

John Bunyan (1628-1688), wrote *The Pilgrim's Progress.* Insightful moral.

Baruch Spinoza (1632-1677) marvelously comprehended his mental, bodily, and external world activities.

John Locke, M.D. (1632-1704) held that all knowledge of life must be provided by the senses.

Gottfried Wilhelm Leibniz (1646-1716) considered mind and body as two independent streams of self-activity.

Abbé Étienne Bonnot de Condillac (1715-1780) developed Locke's idea of the neonatal mind as a *tabula rasa*. He made the profound observation: "Science is a well-made language," featuring the truth that every science is a psycholinguistic production.

Julien Offray de la Mettrie (1709-1751) likened man to a machine, an automaton. Contemporary cybernetics attempts cross-fertilization of sciences by use of interdisciplinary language.

Adam Smith (1723-1790) reduced all economic processes to egoism of the individual, thus indicating, as did Aristotle, that man is by nature a political animal.

Jeremy Bentham (1748-1832) in his *Theory of Fictions* reflected insightful linguistic psychology, "To language, then—to language alone—it is that fictitious entities owe their existence; their impossible, yet indispensable existence."

Pierre Jean Georges Cabanis (1757-1808) divided mind into conscious, semiconscious, and unconscious states. He also founded morality on physiology.

Albrecht von Haller (1708-1777) traced muscular contraction to the muscle itself, thus introducing an innerness of greatest consequence.

Franz Josef Gall (1758-1828) and his associate Johann C. Spurzheim (1776-1832) introduced concepts of localization of cerebral function, beginning with phrenological theories. The idea of such regionalism was strongly resisted by colleagues who considered mentality to be a manifestation of the whole organism, and apparently did not recognize any and all localization or regionalism as consisting entirely of a direction of wholeness. Hence subsequent evidence of justifiable unity of structured functioning continued to be strongly resisted and direly needed. In this context, I mention the 1950 publication of Professor Paul Schilder, a former student of Professor Sigmund Freud, *The Image and Appearance of the Human Body.*

Charles Bell (1774-1842) localized afferent sensory and efferent motor spinal cord roots.

Francois Magendie (1783-1855) furthered Bell's work of elucidating specificity of meaning of neurological structure.

William Hamilton (1788-1856), Scottish genius, claimed the ignoration of self to be the ignoration of one's divinity, "Consciousness may be compared to an internal light, by means of which, and which alone, what passes in the mind is rendered visible."

Marshall Hall (1790-1856) distinguished voluntary and involuntary nervous activity, thereby heeding conscious and unconscious will.

M. J. Pierre Flourens (1794-1867) determined functions (by extirpation of tissue) in several regions of the brain.

William Andrus Alcott (1798-1859) propounded his theory of "Christian Physiology," uniting his science and religion in the effort to attain individual hygienic improvement. Although his motivation was to formulate a physiological rationale for vegetarianism, and although his methods at times seemed symptomatically individualistic, he did emphasize the unity of body, mind, and spirit. He was convinced that "Christian ministers needed a thorough knowledge of physiology to properly understand the scriptures."[4]

Johannes P. Mueller (1801-1858) formulated the doctrine of specific nerve energies, for example, sight and touch. Portentous loçalization of functioning!

Charles Robert Darwin (1809–1882) achieved monumental work in accounting for mind as a growth and development of structured meaning. He identified mental process as evolved biological function serving survival. His writings on emotionality are masterpieces of psychical physiology.

Claude Bernard (1813–1878), insightful physiologist, minded his scientific effort as intraorganismic study. His continued discovery of the organic unity of individuality provided realistic foundation for understanding how behavior works.

Emil du Bois-Reymond (1818–1896) and Herman Ludwig Ferdinand von Helmholtz (1821–1894), each a student of Mueller, greatly furthered his work of exact scientific procedure, including ingenious scientific imagination consciously uniting sensation with meaning (mind).

Herbert Spencer (1820–1903) knowingly cultivated an evolutionary basis for his psychology out of his theory of cerebral localization, "Localization of function is the law of all organization whatever: separateness of duty is universally accompanied with separateness of structure, and it would be marvelous were an exception to exist in the cerebral hemispheres" (*The Principles of Psychology*, 1870–1872).

Mary Baker Eddy (1821–1910) needed to take care of her personal health in a way that she could not find in the medical practice she tried extensively. She gradually and brilliantly concentrated upon the efficacy of *conscious* mental force, *other than that* unconsciously practiced by the medical scientist of her day, ultimately founding her Christian Science. Her accepted concept of mind ruled out a consideration of any so-called materialistic medical psychology. She enjoyed obvious success as a practitioner and teacher of the health benefit derivable from reliance upon subjectivity. Certain pupils, notably Margaret Laird, discovered the health benefit in honoring the absolute inviolability of individual life, in repudiating neither medical nor religious helpfulness.[5]

Paul Broca (1824-1878) discovered evidence of cerebral localization of behavior, supported by autopsy finding, by uniting the capacity for speech with the left cerebral hemisphere, third frontal convolution.

Wilhelm Wundt (1832-1920) is regarded as an esteemed founder of the modern science of psychology. In 1874 his book *Foundations of Physiological Psychology* appeared, featuring the scientific method of empirical research and experiment.

John Hughlings Jackson (1835-1911) succeeded in psychicalizing his neurological findings by observing that activity of specific cerebral areas results in specific psychical activities. He was a consciously original scientist who advanced far beyond his fellow scientist who could not repeat his work because he could not do what Jackson did.

By animal experimentation Gustave Fritsch (1838-1897) and Eduard Hitzig (1838-1907) centralized sources of motor behavior in the dog's cerebral cortex.

Advances in neurosurgery, biochemistry, and psychopharmacology have contributed to understanding the nature of the whole person.

Gustave le Bon (1841-1931) described his world as becoming increasingly dominated by crowds. These collectivities exhibit loss of individual consciousness with the emergence of individual irresponsibility, animal-like mind. Also see Felix Wittmer's *Conquest of the American Mind* (1956).

Ivan P. Pavlov (1849-1936) seemed to develop his conditioned-reflex knowledge in terms of physiology rather than psychology.

Hans Vaihinger (1852-1933) in his *The Philosophy of "As If"* elaborates upon the extent to which false assumption plays a role in man's attempt to understand his experience. I apply this acute conception herein.

Sigmund Freud (1856–1939) found the encompassing psychological framework of his life. His is the genius responsible for realization of the never-ending ideal of man: conscious self-understanding for his inviolable wholeness and innerness. He pioneered a consistent, purely psychological point of view of the whole person that can be appreciated without the need for denying the reality of any meaning whatsoever. His is first place as a scientist of the whole individual. His arduously attained conceptual framework for studying one's own nature is a unique contribution to his fellowman who for untold centuries has been seeking this precise source of conscious self-redemption leading to just life appreciation and fulfillment. I hail my Sigmund Freud as my peerless psychical physiologist.

Charles S. Sherrington (1857–1952), brilliant neurophysiologist, recorded his psychical findings in his renowned book, *The Integrative Action of the Nervous System.* He sensed the value in consciously psychicalizing (making conscious psychology of) physiological findings.

Jacques Loeb (1859–1924) sought to explain behavior in terms of chemistry and physics, conceiving impersonal tropism as motivation.

Otto Jesperson (1860–1943) wrote *The Philosophy of Grammar* (1924). An effort to further needed insight into the innermost nature of human language and of human thought. *Novial* (1928) presents a universal language.

Adolf Meyer (1866–1950), pioneer psychobiologist, originated the term ''ergasiatrics'' to specify man's unitary nature. The term ''mental hygiene'' is also attributed to him.

C. M. Child (1869–1954) created the physiological

gradient concept differentiating metabolic activity from the head end to the tail end of the organism. (Thomas P. Brennan applied this biological orientation to his lectures as professor of psychiatry.)

Walter B. Cannon (1871-1945) studied the intraorganismic meaning of physiology, contributing to understanding of mental stability and emotionality. His appreciation for the psychophysiology of emotion reflects great insight.

Shepard Ivory Franz (1874-1933) and Karl S. Lashley (1890-1958) studied the psychophysiology of learning.

John B. Watson (1878-1958) identified mind as body. The discovery of this unity furthers possible appreciation for the wholeness of individuality. It would provide a new beginning by reducing all of one's activity to the meaning of behavior which appears to be observable as objectivity. Thus it would seem to be able to bypass the necessity for subjectivity of the observer, as such. *As long as the development of the psychologist proceeds along the course of objectivity, it provides consolation. The innerness of life need no longer be an obstacle, once any behavior can seem demonstrably overt.*

Albert Einstein (1879-1952) pointed out in his theory of relativity, "The world of experience as I live it must be *my* creation." He observed that the reproach of subjectivity to the psychologist is applicable to other so-called natural scientists. All sciences consist of psycholinguistic conceptualizations carefully constructed (by each scientist) to be mutually exclusive of each other.

The psychophysiological studies of Samuel T. Orton (1879-1948) have contributed significantly to the understanding of learning.

Scholars found it natural to conjecture how elementary mental power distributed throughout the organism might develop "glorification" in its head end, such as heat spots becoming vision, touch spots becoming hearing, intestinal chemical spots becoming taste and smell, sensibility becoming consciousness, visceral tensions becoming feeling and thinking, and so on.

Edgar Douglas Adrian (1889–   ) created brilliant work in electrophysiology. Hans Berger (1873–1941) studied electrophysiology of the brain initiating electroencephalography. Lee Edward Travis (1896–   ) pursued electrophysiological experimentation especially for its psychological worth. The author collaborated with him.

Professor Walter H. Seegers (1910–   ) teaches integrative physiology indicating that each student's mind grows his own understanding of his physiology and that it is biologically adequate to heed that physiological orientation.

Albert F. Ax (1913–   ) for well over a decade has been achieving outstanding results in his field of psychophysiology, unheralded though they may be. As program director he conducted clinical research in psychophysiology at Lafayette Clinic (Detroit) until its discontinuation in favor of chemical research. Then he moved to the University of Detroit where his psychophysiological program is progressing. Dr. Ax has been a leader in establishing the Society for Psychophysiology Research and the *Journal of Psychophysiology* (some two thousand subscribers). He recorded of the Symposium on Emotion at Loyola University:

> The exciting part of this symposium for me is the discovery by Karl Pribram (and I believe by other neurophysiologists) that they must interpret their neurophysiological findings in psychological concepts like "plans," "images," and "feelings." That psychologists have been saying this for centuries is not nearly as salient as it is for a reluctant neurophysiologist to grudgingly come to this conclusion. This constitutes a breakthrough because it means that finally

really objective facts of neurophysiology will be interpreted as psychic entities. This is what psychophysiology is all about.[6]

I list only a few devoted pioneers in psychosomatic medicine: George Groddeck (1866-1934), Felix Deutsch (1884-1964), Smith Ely Jelliffe (1866-1945), Viktor von Weizsäcker (1886-1957), Flanders Dunbar (1902-1959), Franz Alexander (1891-1964), Thomas M. French (1892-1974), Therese Benedek, Stewart Wolf, Harold Wolff, Jules Masserman, George H. Pollock, Sheldon T. Selesnick.

My study of glorious Homer, Shakespeare, Goethe, Wordsworth, Coleridge, Keats, Blake, Emerson, Thoreau, Melville, Hawthorne, Oliver Wendell Holmes, Whitman, Emily Dickinson, Sidney Lanier, indeed of my every poet, has affirmed my solipsistic self-orientation.

This comprehensive life-orientation, of unlimited conscious responsibility for being all and only one's self, requires a fully developed imagination. My not developing my imagination necessarily carries a grievous health prognosis since it limits the development of my conscious self-identity. My living is the source of all of my creativity, including my creative imagination. Sir William Hamilton speculated of Aristotle that he had an imagination as great as that of Homer. My country's brilliant essayist, Edwin Percy Whipple (1819-1886) said of Sir Isaac Newton (1642-1727) that his imagination "in boundless fertility is second only to Shakespeare's." America's Jonathan Edwards (1703-1758), foremost theologian and President of the College of New Jersey (now Princeton University), united his large understanding with a consecrating imagination devoted to cultivating the moral force of consciously responsible individuality. Benjamin Franklin (1706-1790), ideal American statesman, directed his imagination in interests and research characterized by practicality (worldly wisdom).

Vitally academic John Dewey called Emerson "the one citizen of the New World fit to have his name uttered in the same breath with that of Plato." Divined Emerson, "In all my lectures, I have taught one doctrine, namely, the infinitude of the private man." Of his present age he heeded, "It is said to be the age of the first person singular."

F. O. Matthiessen produced a self-consciousness masterpiece in his monumental work, *American Renaissance,* a self-exciting saga of the historic growth of conscious organic unity.[7]

There follows a brief list of life-insightful authors. Although there is many another one devoted to cultivating appreciation for his (her) own whole life, I wish here to note St. Augustine (354-430), St. Thomas Aquinas (1225-1274), Arthur Collier (1680-1732), George Berkeley (1685-1753), Immanuel Kant (1724-1804), Johann Gottfried von Herder (1744-1803), Johann Heinrich Pestalozzi (1746-1827), Jean Paul F. Richter (1763-1825), Johann Friedrich Herbart (1776-1841), Thomas Jefferson (1743-1826), see his marvelous writings edited by indomitable Julian P. Boyd; Destutt de Tracy (1781-1836), Søren Kierkegaard (1813-1855), Jean Martin Charcot (1825-1893), Thomas H. Huxley (1825-1895), Michel Jules Alfred Breal (1832-1915), Moncure Daniel Conway (1832-1907), Andrew Dickson White (1832-1918), William T. Harris (1835-1909), Franz Brentano (1838-1917).

I mention only a few consciously self-contained authors with scant reference to their works: William James (1842-1910), *Principles of Psychology,* (1890); G. Stanley Hall (1844-1924), *Adolescence* (1907) and *Senescence* (1922); Theodor Lipps (1851-1941), *The Basic Facts of Mental Life,* (1883); J. G. Frazer (1854-1941), *The Golden Bough,* (1890); Sigmund Freud (1856-1939), the standard edition of his complete psychological works by the Hogarth Press and the Institute of Psychoanalysis, (1966-1974); George Thomas White Patrick (1857-1949), his *Introduction to Philosophy*, (1924); Jacques Loeb (1859-1924), *The Organism as a Whole* (1916); John Dewey (1859-1952), in all of his works; Alfred North Whitehead (1861-1947), *Science and the Modern World*, (1925); Adolf Meyer (1866-1950), *Psychobiology: a Science of Man* (1958); William A. White (1870-1937), *Foundations of Psychiatry* (1921); Abram L. Sachar (1899-), *The Course of Our Times, The Men and Events That Shaped the Twentieth Century* (1973); Rupert Emerson (1899-), *From Empire to Nation* (1960); Marquis Childs (1903-), *Eisenhower*, Captive Hero (1958); David Ben-Gurion (1886-), *Israel: Years of Challenge* (1963); Jan Christiaan Smuts (1870-1950), *Holism* (1926); Ernst Cassirer (1874-1945), *An Essay on Man* (1944); Albert Einstein (1879-1952), *The World as I See It* (1934); Werner Jaeger (1888-1961) *Paideia* (1939-1944); Ernst Kretschmer (1888-1964), *Body Build and Character* (1925); Warner Fite (1867-1955) *Individualism* (1911), *Moral Philosophy* (1925); Edward J. Kempf, *Autonomic Functions and the Personality* (1924); Anna Freud (1895-) *The Writings of Anna Freud*, seven volumes, International Universities Press (1966); Kurt Eissler's publications, notably here: *Medical Orthodoxy and the Future of Psychoanalysis* (1965); Karl Abraham (1877-1925), *Selected Papers on Psychoanalysis* (1966); Ruth S. Eissler and Anna Freud, Marianne Kris, and Albert J. Solnit, editors, thirty volumes of *The Psychoanalytic Study of the Child* (1945-1976); Martin Grotjahn, *Beyond Laughter* (1957); William A. Greene, *Hematology and the Derivation of Psychosomatic Concepts* (1975); Mary A. B. Brazier, *The Growth of Concepts Relating to Brain Mechanisms* (1965); Wilder Graves Penfield (1891-), *Excitable Cortex in Conscious Man* (1958);

39

Hans Selye (1907-   ), *The Stress of Life* (1956), *Stress Without Distress* (1974).

Dr. Hans Selye, by growing his much needed concept "nonspecific stress," and by further developing Claude Bernard's elucidation of the "internal environment" and Walter B. Cannon's understanding of "homeostasis," has rendered indispensable health service. He duly respects the biological wisdom in properly heeding the meaning of the wholeness of the individual (particularly his self-understanding) for his preventing, diagnosing, and treating any of his so-called manifestly "regional" health troubles.

For trying to work up the realistic meaning of anyone or anything of my world, I must first renounce *its* every exteriority appearance as being merely illusional externality, and then strive creatively to imagine myself as living only *its* inviolable withinness.

The attempt has been made to set up a kind of spiritual geography, in which the world of human interests is mapped out into kingdoms, this being assigned to philosophy, this to religion, this to science, and so on, each territory separated from the others by defended frontiers. To my mind the problem does not present itself in that form at all . . . Nothing could be plainer than the interpenetration, at every point, of the business in which the three are respectively engaged.

L. P. Jacks, *The Atlantic Monthly,* February, 1924.

THE PSYCHIC NATURE OF PHYSIOLOGY

Repeatedly and repeatably I am finding my physiological processes to be on the scale of my conscious self-perception and traceable in detail. Also I am providing myself with physiological information that I formerly deemed impossible, by teaching myself how to monitor the living of my visceral activity. This kind of so-called "biofeedback" self-knowledge, duly heeded, empowers me to develop volitional control over the behavior of my internal organs, from which I formerly seemed alienated. It enables me to *observe* the physiological consequence of how I behave myself, providing me with a wholly new and unexpected way of "learning from experience," if I will.

This awakening to a life I had heretofore led unconsciously, as if autonomous and involuntary, provides an authoritative revelation of my intact wholeness theretofore only intuitively suspected. My physiology is no longer merely the study of impersonal body functioning but is also the lifesaving study of itself in, and of, me.

Many a seemingly well-informed reader may justly complain. "As St. Paul said to the Romans, 'It is high-time that we awake out of sleep.' Why go to all of the trouble of perpetrating a book extolling the fact, or the necessity for the fact, that you have to use your own mind to produce any set of verbal meanings whatsoever, be it rhetoric, religion, philosophy, science, groupism, or what not? It seems to me that if anything can be safely taken for granted, *that* can be. Why must you labor the obvious truth that you have to 'use your head' in order to accomplish anything at all? What difference can it make if I say, 'I use my brains' or 'I use my mind,' so long as I know what I mean by either one, namely, that I am a panpsychist using my words to describe my understanding of myself?

"Are you implying that you and your colleagues are, as Pope put it, 'harmoniously confused' by your deliberately separated disciplines you call psychology and physiology? Are you unwilling to consider applying what you know about your physiology to what you know about your mind; or what you know about your mind to what you know about your physiology? Certainly, if what you mean by either 'psychology' or 'physiology' is distasteful to you, such is a sure sign of that much of your temporary unreadiness to work yourself up in, or over, either one. After all, whatever you learn to call 'physiology' is just as much your mental living only, as is whatever you learn to call 'psychology,' and if you are tending to use any scientific discipline to disrespect your own life with it, it is lifesaving to discover and take care of that predicament, is it not? Besides, according to you, all knowledge is absolutely incommunicable anyway! In scientific technical terms, what do you call yourself?"

To all of which questioning I bring to notice only, *What a wonderfully gentle method for advancing conscious self-acknowledgment, kind self-questioning can be.* The main purpose of my writing is to help myself. For my reader, may his author's writing prove valuable in disposing him to look for help consciously in the only direction in which he can see, namely, within. I value truth itself in that it helps me to understand and govern the nature and needs of my own wonderful life. The explanation of my physiology, or psychology, cannot be discovered "outside" of it because it lies *inside* it. As Spinoza divined, "Reality is that which explains itself and needs nothing else to explain it."

Technically I might call myself an idiopsychist, considering my mind as the power providing for the function, or meaning, of my wholeness living. *As if* "relationship" of the mind-and-body duality, is overlooked *identity* of their unity. All of my pleasure of living is the consequence of my quickened wholeness vitality; all of my pain or sorrow of living is the consequence of my inhibited wholeness vitality. This strangely individualistic designation, a responsible idiopsychist, applies to the peak of my purposive mental discipline arrived at by hard labor in self-understanding. It indicates further that I also continue to be each kind of psychist I ever was throughout my strenuous psychic development. Except to avoid symptomatic consequences, it is not otherwise necessary for me to so-called psychicalize my psychology, anatomy, physiology, or whatever discipline I have verbalized.

To begin with, my each and every word *is* entirely and only my verbalized mentality, consisting of my own vital being, merely but mightily. I may acknowledge it as entirely my own creation (solipsism) or assert its objectivity (behaviorism).

Yes, I know every other man or thing by, in, and of myself only. The profound divergence of the ideal of recognized self-consciousness from the pandemic forms of my academic psychologist is fully explained by the latter's conscious, or more often unconscious, dependence upon my tempting principle of behaviorism.

45

My self-consciousness consuming delusion of gain or loss of so-called objective possession clouds all appreciation for the spirited growth and freedom of my mind. For instance, my claim to gain advantage for me at the cost of sacrificing *my* fellowman, contradicts itself. My very disclaimer, "not-I," contradicts itself, locating dislocation as it does.

Yes, my knowledge is no more communicable than is my eye or ear. One person can *never* know anything about another person or thing, I am certain. Without a doubt, I cannot overrate the health importance in the study and practice of conscious physiology instead of unconscious physiology.

Every freshman medical student must create in and of his own mind, only, some sixteen-thousand new words. Whether or not he acknowledges his necessity that *he* originate each term, is most important for his appreciating his wholeness. In *as if* leaving my mind to speak of cerebrum or brain I pass from consciously to unconsciously verbalizing my mind. I can create wordage of and about my own mind only. To understand the anatomy or physiology of my body I must invest each of its verbal meanings with my self-interest, either acknowledged or unacknowledged. How enlivening it is to be able to seek beyond my troubled front the full purport and direction of my ever helpful and clearly perfect wholeness! In no other way can I find access to the true working of my mind.

The wholeness of my physiology is all that there can be to it. Its psychic nature is its only nature. All that can be known is the creation of the knower. My physiologist acts his mind from internal motives only, gradually thus cultivating his belief in the truth of himself. Objectivity overlooks vital truth of the whole objectivist, and therefore must be symptom-forming.

Symptom-formation involves conscious self-aliena-
tion, thus obscuring the symptom-former's conscious
self-responsibility and conscious self-control. For example, an hallucinated experience can be felt with greater
intensity and clarity than a sensed or perceived experience.

Although I gladly give my life and thought to conscious
self-wardness, I must ever cope with my habit ("second
nature") of mind drawing my interest to the soft
sweetness of "objectivity." I brought myself up on my
illusion of actually experiencing my world beyond me,
instead of on my actually believing in its provident
existence. Ever since that prior form of self-helpfulness
I have made myself understand it as an early psycholinguistic stage of my development for enabling my later
growing further realization of my inviolably intact individuality. Poet-philosopher-psychologist Samuel T. Coleridge declared,

> I define life as the principle of individuation, or the power
> that unites a given all into a whole which is presupposed
> by its parts.

My personal work on achieving realization of the
necessity of the psychicality of physiology, as of psychicality of psychology, has a long history beginning
in my early imaginings strengthening my imagination
(thus consciously idealizing my mind) and converging
distinctively in my mind-gripping medical school studies
at the University of Iowa. As a medic I could not but
notice the mighty role of morality in the growth of
so-called physical trouble, quite as in the growth of
robust "physical" health. And by *morality* I mean
merely: grateful obedience to the law of my own ever
growing constitution. Yet, it had been my habit of mind

47

to relate moral power strictly to psychical, not including physiological, self-knowledge.

Gradually it became my providence to discover the psychic identity underlying all of my meaningful productivity, body and mind alike, so that I could begin to acknowledge the moral providence pervading my adequate care of my physiological nature and needs. I could understand my George Santayana's dictum: physical integration is a prerequisite to moral integrity. *My experiencing my body or any of my environment is not merely the source of my self-knowledge but is already my knowing mind itself.*

In all of this gratefully observed sketch of my conscious psychogenesis, I bear in mind that it consisted of vicissitudes in my teaching myself *how* to discover the unique reality of myself, namely, my *wholeness nature.* I try to make this understanding dominant.

This "how" proved to be my systematically growing my self-consciousness by a process of growing strong-minded enough to begin to renounce specifically my blind reliance upon my symptomatic conduct of my life. I *must* rely upon my symptomatic behavior for self-help, but it is a great advantage to begin to try to do that with my eyes openly reporting self-perception. Being convinced myself of any truth relieves me of the compulsion to try (vainly) to convince someone else of it. With my St. Augustine I can keep my peace, "Let others wrangle, I will wonder" (*Alii disputent, ego mirabor*). For, ever since I could renounce my habit of taking-for-granted the marvelous nature of my being, I have instead exercised my appreciation for its wonderfulness in my work and felt this exercise as giving a lifesaving development to my wholeness.

I try to write in the light, as for my own reading only, and I aim at making every passage of my life equally self-continent. I am sure that my mightiest contribution to myself in the name of my fellowman is my cultivating my capacity for minding my own business, for strictly attending to all that can concern me, namely, my own welfare that includes *my* fellowman's welfare. I must ever be translating "What will people think?" into *How am I thinking my own thought.* Thus I conceive the heavy duty that belongs to my living of my fellowman.

A most difficult moral is that of standing by silently and determinedly letting my conscious self continence, rather than my sore side, speak for me. I discover that man is so created: the meaning of all of us inhabits each of us. The more social the man, the stronger is his individuality. The conqueror of men is uncontrolled in himself. When I defer to my fellowman I defer to his perfection as a whole individual. My great Calvinist luminary, Henry Scougal, sermonized, "Learn to adore your nature." The wholeness of it, certainly.

I have found that the more I can acknowledge being all of my own living of my experience, sorrowing and rejoicing in my self alone, the stronger minded I become. In other words, it is my laboriously attained appreciation for my wholeness that necessitates my realization that my so-called physiology is as psychic as is my philosophy, biology, chemistry, or whatever meaning, including psychology itself.

The identical psychicality of mind-and-body is revealed by the translation of physiological (body) language into psychical (wholeness) language, or the converse (namely, the translation of psychical into physiological language). If this short screed can amount to a contribution favoring renunciation of prejudice against considering the immeasurable power of mind, I am satisfied with it. Not long ago addressing my medical colleagues

I mentioned evidence of the identity of allergic and psychic symptom formation. Certain physicians seemed ready for it.

For example, I mentioned this striking similarity. To produce an allergic reaction the antigen, *created by the patient,* must first be present to make the allergen effective. To produce a psychiatric symptom the resistance, *created by the patient,* must first be present to enable the mind to repress any of its experience. Then, the return of the repressed mentality in a disguised form constitutes the symptom.

I felt keenly that I could not have chosen a less popular theme for my address than that of associating psychic experience with allergy.

As a materialist, an objective physiologist, my mind-blindness spares my assuming conscious responsibility for living my own experience at the cost of that much conscious self-belittlement. My choice is to continue growing myself or die. Growing myself means finding room for my every new development, a process that is difficult and hence regularly resisted. Thus it happens that the kind of person I am can resist seeing its identity in the kind of person I can become, regardless of how beneficial my new growth may prove itself to be. As an objective scientist I do not come into harmony with self-truth, the mental process essential for establishing the realization of my conscious wholeness.

My physiology professor, John T. McClintock, uniquely presented not only the terms but the meanings called sensation and perception as psychical in nature. My ethics professor, E. D. Starbuck, understood the organic goodness of life. My psychiatry professor, Samuel T. Orton, was exclusively devoted to understanding the mind from a strictly organic point of view. Particularly when I began my psychiatric residency with him,

I became necessarily knowledgeable regarding the startling truth of cerebral localization, a worthy sunburst enlightening my understanding of my psychology as mental physiology. My worthy psychology professor, Carl Emil Seashore, seemed to favor applied psychology.

A scientist of powerful understanding, Professor Orton was duly skeptical of much of the dynamic psychiatry of his day. Thus, he looked for the only source of mental trouble solely and wholly in the individual concerned, and of that one his examination was exceptionally thorough. Since the psychodynamics of his day seemed to relate mental trouble to disturbance going on *between* the patient and somebody else, such as his parent or sibling, Professor Orton did not encourage the study of it. Thus he honored the basis of medical research and treatment, namely, the *individual variant*. His organic viewpoint was well grounded in his self-identifying study of scientists who advanced psychological understanding by creating new ideas of the action of the nervous system (Franz Josef Gall, Herbert Spencer, Brown-Sequard, Charles Bell, Thomas Laycock, John Hughlings Jackson, R. Cajal, Charles Sherrington).

My immediate supervisor, George S. Sprague, was at first an overwhelming experience on account of his profound understanding of the subjectivity of his individuality. His patient approval of his trainee's self-unconsciousness, "honor rooted in dishonor" (Tennyson), was most helpful. With him I practiced my first free associating, confronting myself with my mind's areas of painfulness that prevented my consciously comprehending the functioning of my wholeness. Professor A. D. Ritchie sensibly records, "The organism is the way it behaves and it behaves as a whole" (*Natural History of Mind,* p. 184).

My mind can and does hypostatize "independent

entities" such as "objects," "representations," "relationships," and similar meanings *of its own production,* supporting a so-called commonsense vocabulary. Thus it is, I can use my mind as if some of it is not mine, is impersonal, is not mental, and so on. It is the office of the psychologist to study how his mind can thus modify itself in the interest of the functioning of its protean nature.

I can develop conscious purpose for the directing of my own self-growth as my conscious self-identity augments itself. A consciously whole-formed mind reveals itself in strong, independent character of self-mastery organized to observe and obey the best interests of the whole individual. It uses itself to personalize, not depersonalize; to internalize, not externalize; to organize, not disorganize. It enjoys its conscious integrity, being of one piece, being at one with itself, being a consciously intact mentality, being the central synthetic power of all of its meaningfulness.

My years of psychological research work with Professor Lee Edward Travis in electrophysiology further enabled my ability to conceive all of my study of physiology as being psychical in nature. I dared even to imagine the possibility that my study of action currents in my living patient might reveal findings translatable into a vocabulary verbalizing his mentation. Certainly I considered that the concept of electrical energy might lead to understanding energy in dynamic mental process. Sherrington had recorded: "Electrical potentials indicate nerve activity closely and quickly. The nerve impulse . . . seems in essence electrical."

My Professor Albert M. Barrett's open-mindedness vitalized my conscious freedom to follow my mind's growing of its so-called psychiatric experience. He purposefully pursued the ideal of total research and training, a perspective specifically conducive to my cultivating self-understanding. At the University of Michigan I also had helpful psychological discussion with Dr. Theophile Raphael, the head of the university student mental health service and a former First Assistant of Professor Albert M. Barrett, concerning some of those great things for which literature was too small "and only life large enough" (Chesterton).

With Professor Barrett's help I secured the unique opportunity for an extended sabbatical leave and, ultimately, the privileged work with my Professor Sigmund Freud.

My physiological study can be nothing but my learning about a circumscribed area of my own mind. Whether or not I heed that truth of my own living is of lifesaving consequence. Merely my specifying certain sets of words to signify particular life interests of mine does not, and cannot, make them *as if* heterogeneous segments of my homogeneous mind.

Happy is the man whose nature sorts with his vocation, wisely observed Francis Bacon. Healthfully happy is the man whose nature *consciously* sorts with his vocation. Learning without self consciousness is enslaving labor.

Since my every learning discipline, including science, is a development of unified meaning in a specific area of my own homogeneous mind, the condition for recognizing the basic identity (my own self's identity) underlying all of this differentiation is already given. The all-important distinction remains however: to be "given," is not the same as to be *consciously given*. The illusion that one life interest of mine can be more or less the product of my own mind than is another, is tenderly cherished and therefore needs to be tenderly renounced. *Renounced* it must be however, if I would protect my health, for self repudiation is the innocent source of all of my belittlement of my own life. In this writing, by "renounced" I do not mean rejected as repressed, but quite the contrary. By renunciation I mean: recognizing limited helpfulness in any meaning so that I can at will lovingly extend that helpfulness.

By *psychicality* of physiology I mean: 1) my mind's acknowledgeable study of all of its meaning that I classify as physiological, and hence 2) my full recognition and cultivation of the convincing evidence that all of the meaning of my physiology *is* mental. However, I freely acknowledge that my ability to formulate this holistic perspective on physiology has been the result of my arduously working up a new way of minding my living and of living of my mind. Historically, though, I revive only Aristotle's complete assurance that the body and its thinking are only one existence (*De Anima*).

All of my hope for biologically adequate self-fulfillment hinges upon the way I teach myself to use my mind. When I turn from using it *admittedly* I thereby forfeit my awareness for the only possible truth about its activity, namely, the functional nature of my own behavior. Thereby I create for myself the innumerable and insuperable phantom problems based upon duality or plurality of any degree: mind-body, spirit-substance, real-imaginary, internal-external, divine-profane, unity-division, innate-acquired, end-means, right-wrong, good-bad, gain-loss, success-failure, yes-no, love-hate, now-then, here-there, and all such illusional double entity traceable to the unacknowledged whole-integral one.

As long as I refuse to acknowledge personal responsibility for being all of the whole of my own living of my experience, such as my sensations, percepts, ideas or whatever I live, I thereby impose upon myself an (illusional) impassable remoteness between my question and answer, each of which really consists of my own identity and, as such, lends itself ideally to my study of all that there can be for me to study, namely, my own nature.

Dean William R. Inge writes: "I want . . . finally to inquire whether any signs of unitary purpose can be found in History" (*God and the Astronomers*, p. 125). Nothing but unitary purpose can be found in my history of my own psychogenesis. One is all and only about itself. I can have nothing further to learn from, other than from my growing my own further experience of me.

Merciful truth; remorseful fact! Only because of immersing myself in my own illusion of externality can I suffer rather than enjoy my life. From early childhood on, my unconscious habit of mind is to turn my only possible life-orientation (which is self-growth) completely around so that I honor my imagining the existence of externality at the cost of disregarding my only directly observable reality, namely, my subjective mind. And I achieve this mental deception by depreciating my innate subjectivity as being only "imaginary," so that I can exalt my imagined externality as being real.

My only possible experience must be personally mine. *I only* can create, out of my own living, *all* of my world, and then can imagine it to be only itself, other-than-me, *as if* outside of me. Thus I, including my every fellow-man, try to avoid having to accept *on faith only* the existence of my external world. My mind alone creates concepts of materials such as atoms, chair, soil, man, universe, and so on. In addition I can accept on faith alone the existence of anything I mean by *external world*.

I can discover truth for myself only, and it must be of myself only. The activities of my mind are my only possible sources for observation, and of observation.

All that I can ever sense or feel or know pertains only to my own mental functioning. The physiology of my so-called body, including my nervous system, consists wholly of psychic observation, psychic research, psychic experiment, psychic theory.

How can there be any overwhelming concern such as, how to "relate" my physiological findings to my psychical findings, since they are already psychical? My sensorium creates all of its sensation out of my own living. My Bios and Logos are one; each is its own other.

Providently, my limited ability to observe and heed the intact *organicity* of any of my life experience requires symptoms signalizing my limitation. Hence it is of first importance that I fully acknowledge 1) the organicity of all of my mental activity and 2) the mentality of my every meaningful experience, shallow or deep. There can be no just question about how to get my mind into my physiology, because it is already there. The needed question is, how did my physiology ever seem to be able to get out of my psychical orientation in the first place? Nothing can be biological that does not exist in the given individual.

I do well to confront each most relevant concern, namely, why and how I 1) as a physiologist repress my psychical subjectivity and 2) as a psychologist repress my physiological subjectivity. Localization of my function at the only place where it is possible for it to occur (i.e., within my individuality) requires heedfulness for my inviolable wholeness as an individual organism. Research upon any aspect or region of my human being can contribute to my understanding of the working of my mind.

All of my research upon my so-called external world is really and merely upon my mental power, and can further contribute to my understanding of the working of my mind. My every astronomical sensation, perception, or computation is nothing but my own mental activity. My so-called sociological research similarly consists completely of my own mental activity. Even my study of religion is wholly a special dimension of study of myself. Whatever way I use my life contributes to my understanding of my life only.

Conscious functioning of my wholeness-nature, itself, is functioning of my only organ of recognizable integration. Every region of my existence is an extension of my holistic meaningfulness consisting of psychic functioning that is either conscious or unconscious. And unconsciousness is the assumed psychical substratum of consciousness.

It is the identity of homogeneous wholeness which is needed to reduce all seeming divisiveness or regionalism to its essence as wholeness. By seeming to be able to divide myself up to conquer what I divide, I end up ignoring my original indivisibility, that is, my inviolable individuality. Thus I can begin to use my mind as if some of it were not mental. I can study my physiology as if it were not my mental physiology, but rather as if it were even *impersonal.*

This deeply significant truth of my wholeness requires sustained heed. What do I mean when I say that *my* everybody and everything is entirely and only mine, consisting of nothing but my own self possession and contributing only to my conscious or unconscious self identity? I mean that *all* of the meaning of my life is created by my mental experience. Thus, from developing my mind I gradually develop a certain unity of meaning which I grow to name "I" and that augmenting psychic complex is invested with the feeling of selfness that I name "personal." I have no difficulty realizing that it concerns me only, that it is all and only mine. In fact, I give it my signature.

However as my mind develops itself it also experiences additional mental unities which I grow to recognize by names other than my own, carefully keeping each such seemingly not-I unity in its own special category of meaning. Nevertheless every one of these unities designating my meaning for my fellow being is integral to my wholeness-nature, even though I maintain it by habit (second nature) as an entity distinct from that to which I give my personal name. In other words, my meaning designating my fellow being exists right alongside of my meaning designating my freely acknowledgeable being, thus:

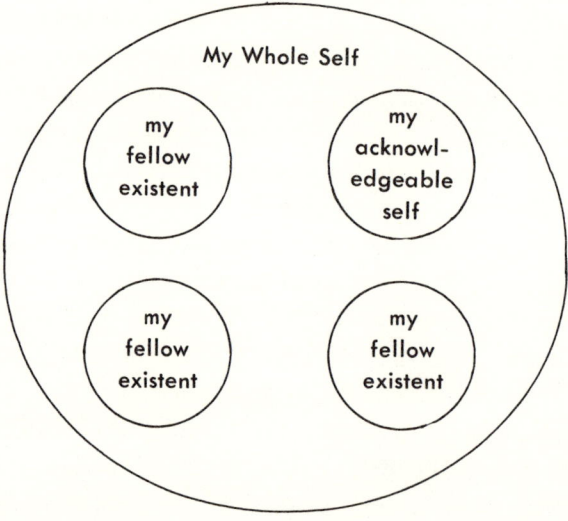

It is possible also for me to imagine my fellow being as existing within my acknowledgeable self as integral to it, and as creating his (her) Dorsey meaning. This latter psychogenesis, that of recognizing my personal identity in *my* fellow existent (including his individuations) enables my cultivation of *my appreciation* for my conscious wholeness, a growth tantamount to my appreciation for my life itself.

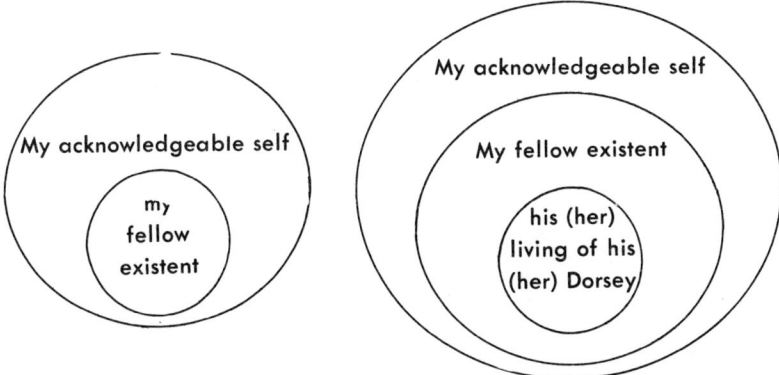

As soon as there is life, there is self-direction,
and absorbing and using of material.

Ralph Waldo Emerson

FURTHER CONSIDERATIONS OF THE PSYCHICALITY OF PHYSIOLOGY

The key to the understanding of the appropriateness of a *conscious* pyschicality of physiology is to be found only in full appreciation for the absolute individuality of the individual. "Only," since such comprehensive heeding of my oneness empowers my understanding that *all* of my being always exists homogeneously, despite the helpfulness I find in distinctifying elements in it as if they might be separable. Thus, all that I can mean by "psychical" and all that I can mean by "physiological" are not merely connected in me somehow, but rather are already my organically united mentality, each "all" being integral to my generative wholeness.

That acknowledged *indivisible* wholeness is the prerequisite of unity, must be thoroughly appreciated, for me to comprehend, as did magnanimously democratic Ralph Waldo Emerson, that *all union is within whatever is united*, that oneness of unity is the ideality identifying the actual meaning of the term *individual*. However this unity may be conscious (rarely) or unconscious (regularly), and right there is the rub.

This cardinal principle, namely, the fundamental truth of identity, or sameness, existing throughout all individuality, must be kept *consciously* functioning in mind if the illusion of its absence created by insightless analysis

63

is to be resolved. By "insightless analysis" I mean specifically a self-blinded effort to understand individuality by seemingly being able to take it apart (divide it up) for the purpose of studying its real nature *which is specifically that of absolute indivisibility only.* Analysis that does not consist of abstracted synthesis is entirely baseless, except in omnipotent imagination. In psychoanalysis the process of my *conscious* synthesis is quietly but continuously creating awareness for my *conscious* wholeness.

Insightful analysis supposes the revelation of synthesis of which its own analyzed elements must consist. A whole is usually said to consist of so-called *parts*, with the result that the necessary wholeness of each "part" may be, and usually is, completely ignored. Every element of a whole can consist of nothing but piecemeal wholeness itself. Piecemeal or regional consciousness is ordinarily unrecognizable whole-self consciousness. Any and every form and function of life, such as my discovering that my limb is integral to my body, must be produced by the living (growing) of it.

The true (comprehensive) definition of individuality is available and accessible *only* to the acknowledged self-scholar who can teach himself how to grow this enlarging conception of his own wholeness for his own life satisfaction. The illusion that one force or entity of whatever nature can have anything whatsoever to do with another force or entity, underlies all rejection of the wholeness nature of individuality.

In my research on physiology as consisting of advancing psychical construction, I repeatedly witness the biological basis of this formulation by observing the process of psychogenesis while it is occurring in my stream of consciousness. Frequently this development occurs in my sleep and I awaken to find its presence. I have learned to get out of bed and immediately write down whatever extended observation of this organic growth is presenting itself, otherwise it may not consciously recur for days or weeks.

My vigilant watching consists of heeding new mentation akin to the theme I am working, whenever it does appear psycholinguistically, awake or asleep, empowering my understanding of the basic organicity of my mental process. It is this kind of living that reveals me to myself as the architect of my own meaningfulness, originator of my own psychic reality, conscious and unconscious. It lies in the power of my conscious wholeness to liberate my will "from the sheaths and clogs it has outgrown" (Emerson).

By associating any of my living with undesirability, I thereby tend to exclude it from the unity of my mind that I count as being my self-identity, thus obscuring my personal view of that much of my actual wholeness. Hatred, not being recognizable as all and only about itself, creates the most of this seeming of not-self. Wisdom is *conscious* self-organization. As already mentioned, to associate any feeling of unpleasantness with any meaning of my mind other than itself alone, is the origin of repression (that is, of alienation of any of my mentality from my acknowledgeable self identity), the source of all of my life depreciation.

Only my reader who can consciously author the above paragraphs can recognize that no science can exist other than that created by the *mind* of an individual man, that man's every science is his *psychic* construction, and that any one of his sciences (including physiology) is just as pychical as is any other one. It is my mind functioning whether or not I responsibly use it for my conscious purpose of studying its protean nature. The fact remains that I have no other possibility. My only choice is to disavow this necessity or to, so to speak, "get with it." My conscious value of myself to myself depends upon my extending the bounds of my appreciable self-identity (my self-acknowledgeable wholeness).

There can never be anything for me to resist, or try to resist (be unwilling to consider kindly), except my very own unconscious selfness. My recognizing all of my behavior as my growing of my life, all and only, reveals my resisting my growing of my self-consciousness as amounting to my inhibiting my mental development. Augmenting self-consciousness results from adding to the previous limit of self-consciousness that it supersedes. The end of my idiopsychology is to detect my unity, that is, my inviolable wholeness. It is my power of self-consciousness *specifically* that affords me a knowledge of the true extent of the wholeness of my life. What I mean by my will power, my conscious volition, is the product of my conscious wholeness only.

Ignorance of my self is my only ignorance of reality. Acknowledgeable idiopsychology is preeminently the science of self achieved by the discipline of self-insight. St. Augustine discovered pain to be the signal of inhibited appreciation for individuality, and pleasure to be the sign of conscious formation of one's unity.

It is indeed well to discover that for growing my conscious originality and independence of my mind I am as much at the mercy of my scepticism as of my credulity. I cannot deny the right to exist of any possible meaning without denying the right of that much of myself to exist. But what is more to the point of wholeness, all of my denial, dispute, or argument merely implies that I am not even discussing the same subject as is my disputant. Basically he is discussing his mentality only, whereas all I can ever do is discuss my own self meanings. A clear concept of inviolable self wholeness must be the basis of all true edification.

Certainly it is true that my mighty force of linguistic

habit is dead set against my conducting my individual life consciously as such. My language is always serving my conscious purpose and thereby deserves all the credit I attribute to it, despite the extent to which it may be holding me to the life of a marionette, hardly ever sensing dimly that I am also the only possible one who can ever pull any of the strings. However I, alone, am the only possible source of my resistance to consciously subjecting my very own language to the overall truth of my absolute wholeness and its implied sameness, or identity.[1]

Most essential for me as a physiologist is my *thorough* understanding that I cannot have any immediate knowledge of anything beyond the modifications of my own mind, although I may be accustomed to seem to be able to divide my indivisible mind into self and not-self meanings. To illustrate, it is most helpful for me to be able to realize that all of my sensations, perceptions, thoughts, and feelings that I relegate to (illusory) relationships with my parent, sibling, spouse, child, or neighbor are actually and only internal workings of my own psychophysiological economy. Thus, my parent trouble, spouse trouble, child trouble or whatever seemingly not-I trouble, is no different from heart, stomach or liver trouble as far as their consisting entirely of my inhibited working of my own individual nature is concerned. My acknowledging back trouble or neighbor trouble is equally and only my observing an objectionable (repressed) fact of my self-continent wholeness which is (helpfully) necessary when it occurs.

My present writing on psychology, physiology, geology, or any other science, must be a conscious exposition of my internal vitality. Thus I can live my world of observable reality, understanding all of my experience as modes of my own being revealing my manner of existing.

Whatever I perceive is only a modification, or mode, grown by my percipient subjectivity. Once I begin to look to my self consciousness for my life orientation, I find I can anticipate steady increase in my appreciation for my intact organic integrity.

The reasoner can argue, indeed must argue, for his mind judges that one meaning or idea can have something to do with another (the *sine qua non* of dispute). The seer sees no advantage in argument, persuasion or the like performance based upon the illusion of plurality, not on the consciousness for oneness. I might as well regard my body (nuclear mentality) to be naught but "meat and drink" as regard my societal mentality to be "acquired relationships."

I am the subject of all of my objectivity; the *as if* object of all of my subjectivity. Like every other duality the subject-object ambivalence is vindicated by the fact that it provides the roots for the growth of the unitary conception of self-continent individuality. Diarizes Aristotle, "Knowledge of opposites is one." Nothing can be understandable except in terms of the history of its own development. The understanding of unity is elucidated by the discovery of the illusory nature of divisibility.

Any existent can attest its own being only. I am capable of appreciating this insightful life orientation just to the extent that I can recognize that I can attest my own self-continent existence only. My existence infers my self-knowledge quite as my self-knowledge infers my existence. Acknowledged self living is the primary affirmation of my individual existence. I count this life development as the greatest discovery in my science of mind, namely, that all of my mind is native to my mind. All of my truth is the creation of my self growth. My will is the production of my mental energy. *My inability to conceive all of my existence as my growth of it, forces me to create the necessity for some other conception of causation.*

It is of lifesaving importance for me to be able to sense and feel my own self identity in all of my self discovery that I label physiology, theology, psychology, geography, or whatever. And not until I can acknowledge my self identity in whatever I experience (studied or played) can my mind rest easy from its innumerable phantom problems. I created each of these troubles (traceable to illusion) by my mind's confirmed habit of seeming to be able to live 1) meaning that is consciously native, and 2) meaning that seems consciously alien (to its currently developed fund of conscious self identity).

The question of whether the mind exists in the body or whether the body exists in the mind, poses exactly such a phantom problem. All that can possibly be meant by body must exist only in its special meaning (mentality); likewise, all that can possibly be meant by mind must exist only in its special meaning (mentality).

Health trouble originates whenever the truth of the inviolable integrity of my organic wholeness is not fully honored, whether I revere what I mean by *body* at the cost of ignoring or discounting what I mean by *mind*, or whether I revere what I mean by *mind* at the cost of ignoring or discounting what I mean by *body*. All conscious negation or denial of the truth of my nature must become unconscious affirmation of it, expressed symptomatically.

We carry with us the wonders we seek without us: there is all Africa and her prodigies in us; we are that bold and adventurous piece of Nature, which he that studies wisely learns in a compendium what others labor at in a divided piece and endless volume.

Sir Thomas Browne. *Religio Medici.*

THE LANGUAGE OF PSYCHICALITY

My conventional linguistic conceptualizations ana categories of psychology and physiology are carefully and consistently constructed to be mutually exclusive, thus preventing any conscious verbal unification, although there may be a strongly felt need for appreciating the underlying identity of the vital functioning of my mind supporting each class of words. My conventional definitions, instead of being reliably based upon historical development, are frequently arbitrary, thus approximating word magic, unfit for the sober language of conscious individuality.

Hence, my invention of my category of psychophysiology offers me the very finest predictive benefits of nominalism as far as my developing a science of the whole self (including its environmental experience as organic) is concerned. Its chief achievement is that of providing a way by which all of my world can be identified as really mine. This powerfully needed help enables me to use my productive imagination for conceiving my universe (including my meaning of me and of all beyond me). I can create only in and of my own image and likeness. Too much stress cannot be placed upon this necessity: all of my scientific effort such as physiology, or physics, or whatever, reflects the degree to which I can be a consciously whole scientist. However the historical fact that nearly all of my primary vocabulary consisted of words naming my own mental experiences *as if* they were not my own mental experiences, easily accounts for the difficulty confronting me if I become so fortunate as to try to create a vocabulary that translates all of my illusional non-mental words into their only possible real meaning as referring entirely to my own mind's creativity.

Such an attempt, translating into my conscious idiolect all of my vocabulary supporting my mind's dissociating itself into acknowledgeable and unacknowledgeable personal identity, is too trying to undertake unless I can bring myself to see the indescribably wonderful health benefit deriving from it. I cannot conceive how deceptive is my use of language unconsciously accommodated to the necessities of my infancy, childhood, and adolescence. That it is enormous is beyond any doubt. That it does function warningly as symptom formation I am convinced.

The role of habit in my mental activity is most deserving of careful heed, for its consequence in keeping my unconscious mentality unconscious is tremendous. For example, accurate measure of how much my mind is needing the conscious functioning of any of its meaning, is the degree of conscious resistance to that functioning that I can *feel*. My "taking for granted" that *habit* of resistance then prevents my use of that resisted meaning for furthering my present acknowledgeable identification of my own experience.

Applying this understanding to my lifelong habit, of overlooking that my purely mental functioning provides my living with any and all of its meaningful experience, I can begin to appreciate my necessity to renounce such a costly custom. "Costly" in that it preserves my sleep for the truth that my psychic reality is my only possible meaningful reality of my physiology, anatomy, geography, or any science.

Growing my physiological sensations and perceptions of my so-called body, as of other *as if* external events, is clear, demonstrable, and repeatable development occurring only in my own mind. *As if* externality happenings, immediately can stir me up or settle me down. They are most notable for their influencing my other biological activities. They can reduce my nature to a state of shock or affect my constitutional status in innumerable ways of life-and-death concern.

Even my reading a book or watching television can affect my mental status to such a degree that I feel I must turn from it or toward it. Children are allowed to experience exciting events of overwhelming tension on the assumption that they are "make-believe externalities," whereas they are really *direct mind traumatizing internalities*. The mind of a child is hardly, if at all, capable of sharply distinguishing his very own psychic reality from his consciously imagined world that exists apart from him and of which he is an integral individual. The same limitation seems to apply to my nearly every adult as well, in my opinion.

It is a well-known fact that (physiological) sensation and perception are of greatest quantitative and qualitative importance for mental health. Too much or too little of either expresses itself as grievous mental trauma.

Just how my mind's experiencing its sensory and perceptual living enters into the physiology of my vision audition, gustation, olfaction, and every other action of my being, remains to be worked up. That it is nearly a totally unexplored region of psychophysiology, however, is in my studied opinion beyond any question whatsoever.

My life consists entirely and only of continuous

modification of my own wholeness. However it often *seems* that my own living is conducted in the midst of so-called surroundings, so that the all-important truth is obscured that I only am living all of whatever *appears* to surround me.

My reader asks, "What do you mean by 'psychical?' And, What do you mean by 'mean?' "

Psychical is mental. To mean is to mind. Such words referring to mentality are not common,—a notable sign that they have not been considered sufficiently important to warrant many synonyms. The abiding fact is that my *every* word is a synonym for my mentality, a name for my self, since it has all of its origin and course of development as my meaning. Again, meaning is mental functioning.

Meaning is a term of greatest consequence, the word of words. Sigmund Freud described it as significance, intention, tendency, and position in a mental series, no less. Meaning is the *sine qua non* of mentality, constituting the whole mind. All that I can ever mean must be about, and of, my own mind. Whatever I experience must be all and only of, and about, me.

For deciding what (psychical) scientific discipline is most suitable for understanding the whole condition of a person, it is necessary to find the one that can include all of the others. My psychicality can subsume the comprehensive range of my biological, physiological, genetic, linguistic, aesthetic, educational, technological, literary, poetic, philosophical, semantic, fictional, economic, political, social, religious, and other living.

My idiopsychology is my one science evidently most capable of widest application and of creating an organic whole out of all of the dissimilar disciplines of my academic and non-academic living. Any and all of my behavior makes psychical sense in terms of my wholeness. Potentially my idiopsychology can organize my complete vitality on every level of integration of its identity. Much of my science is already made consciously mental by me (although I may resist that fact), such as the physiology of the vegetative and central nervous system, endocrine glands, viscera, skin, and the like *as if* non-psychical meanings. Therefore, already all of such psychic reality exists as ready-made linguistic meaning in and of my life. All I need to do is discover that it *is* mine and, if I wish, translate each of my languages into the other.

I must make all of my being gradually become *consciously meaningful* before I can have any notion or understanding of it at all. That means that I must create my being anew as psychical being before I can even observe its existence as such. This is the observable and repeatable process of psychogenesis (mental development) fully validating the claim that mind exists as the marvelous entity it is.

Nothing but mind, itself, as the entity consisting of life's creative meaning, can account for the magnificent working of all that can be named nature, or whatever is distinguishable in living experience. Each of the various theories invented for explaining my behavior is a mental abstraction from mental fact itself (e.g., cerebral localization, behaviorism, parallelism, materialism, pragmatism, theism, etc.).

My conventional psychology can have no recourse to physiology, biochemistry, biology, or whatever when each unit is treated as if each were non-psychical. Rarely may I consciously credit my mind with all of its power of creating my meaning for all of *my* world, but it is essential for its asymptomatic growth and development that I teach myself to do so consistently and steadily.

Disciplining of my mind with responsibility and authority for being all of itself, is the specific meaning of mental hygiene. My mentality is the only active ground of *all* of its activities. All of my experience must originate and terminate in the constructive action of my organic being according to the wholeness principle immanent in it.

My mind attains to its knowledge and systematized experience of every description entirely in, and according to, its own nature, call it by the name of whatever doctrine I will. I choose to call it: *consciously lived solipsistic idealism.* My mind is its own meaning only. How to understand myself as being my only understander, is a statement of the issue that describes all of my psychology as idiopsychology.

Although I can readily imagine that (whatever I mean by) my *living* must be the foundation of my meaning, nevertheless I can accomplish any and all of my imagining only by using my mind for that definite purpose. Whatever the merit of so-called physiological work, it presupposes nothing but pure idiopsychology. Knowledge of any kind is not possible except by way of the mental process of the knower.

My insightless physiological construction of psychology is a contradiction in terms based upon imagining mental facts as if they are not mental and then mistaking them for their seeming non-mentality. Thus I must use my mind to conceive all that I can possibly mean by form, position, solidity, aggregation, motion, space, time, etc. Then, as long as I can blindly pretend that these mental creations of mine are not entirely mental, I can assert that the attributes of matter cannot be ascribed to mind "without absurdity." My illusion of materiality may attain its most convincing force in my dream or delirium.

My reader says, "I have many experiences that I do not want to claim as my own. I find fault, and am discontented, with them. I do not want to experience religious physiology, or moral biochemistry, or concrete ideality, or metallic spirit."

My discontent and my faultfinding are also natural mentality of mine, signifying that I am not yet ready to consider kindly whatever I can now mind only with displeasure. Out of my present discontent and faultfinding can I gradually grow appreciation for my inviolable biological wholeness also. Meanwhile whatever experience I cannot presently prize as integral to my conscious unity necessitates my coping with my illusion of duality, namely: I and not-I.

My hardest lesson is to keep reminding myself that my faultfinding *is* based upon some kind of *as if* duality, whereas my perfection finding is based upon the real unity of being. The moment I engage myself in living as if I can be "out of my mind" I start to become involved in 1) all of the ethical problems traceable to my false position (such as sin, evil, untruth, guilt), and 2) all of the physiological troubles traceable to psychical imbalance (such as sensory or motor complaints).

Liberation of my self from my self ignorance is the end and aim of all of my education and research. My success consists in my steadfast application to this source of my health and strength. Emerson said it beautifully. "The visible carriage or action of the individual, as resulting from his organization and his will combined, we call manners. What are they but thought entering the hands and feet, controlling the movements of the body, the speech and behavior?"

My reader speaks up, "Trying to understand my reading as being my authoring of your thesis, it appears that I must now renounce each erstwhile dominant theory, such as that of psychophysical parallelism, for instance, in favor of conceiving all that I have previously considered as physical to be just as mental as all that

I have previously considered as purely mental.

"To illustrate, when I say 'The mind acts through the nervous system,' all I can be really meaning is: 'The mind acts through the mind, or the *mental* nervous system, if you will.' Similarly, by 'The cerebral localization of mental functioning,' all I can really be meaning is: 'The mental localization of mental functioning.' Or all that I can mean by 'Manifestation of mind is enabled by the aid of material instrumentation' is: 'Manifestation of mind is enabled only by the aid of mental functioning.' As did Aristotle you assert, knowledge of mental and material is one. However it may seem, the protean power of the mind is such that underlying all apparent difference of meaning is its absolute identity.

"If all of that shocking information is true then how must that affect my lifelong understanding of the duality of mind and body as of two opposites? Must I get rid of all such divisions as spirit-substance, organic-functional, subject-object, physiological-pathological or success-error?"

Such earnest questioning in myself leads to my recognizing that *my present understanding is actually the very issue of all of my previous understanding*, that all of my present development is the *outgrowth* of all of my former development, that I depend upon all of my preceding perspective as being the very root of my present perspective. Thus it behooves me to honor and cherish, rather than try to rid myself of, whatever earlier life-orientation I may have created for myself, because it provided me with the indispensable basis for my later appreciation for my living. So I keep on growing and developing.

Due to my repudiating (rather than understanding as necessarily helpful) any of my foregoing way of living, I involve myself in all of the symptomatic consequences of any other kind of repression and, without realizing

80

it, I try to deprive myself of a conscious foundation for my present way of life. Thus my scientific development must continue as if without a history in my personal existence (in my conscious self identity). To be without a history of being, is my definition of magical development.

In all of my scientific research it has been consciousness for my so-called error that has enabled me to *continue* my inquiry in the true direction. However I do realize distinctly how difficult it is to attribute this indispensable successfulness to error at the moment I discover it. Thus I tend to set up the illusional duality success-error without duly crediting conscious error with its essential helpfulness.

As already stated, my prized ideal as a scientist is that each kind of science must be respected as being all and only about itself. To that end I make a place for it in a special area of my mind, to be distinguishable from any other area of my mind. To this end I develop a special language for verbalizing its meanings, and take care not to confuse, for example, physiological and psychological terms, lest I unwittingly substitute a physiological for a psychological description or meaning. Accurate conscious localization of the language of each scientific interest may be duly stressed, but the truth that each localization is mental only may be regularly repressed, either *as if* "taken for granted," "beside the point," "irrelevant," "misleading" or even "impossible."

I practice specific respect for, and reliance upon, my psyche, or mind, as the substance of my every other study, for example, as constituting totally my every meaning designated "a physical condition." In so doing I define 1) the discipline of idiopsychology as the study of a given mind by that mind, and 2) every other discipline as applied psychicality, that is, a given mind's study of a specific set of its own meanings. All of my study of physiology thus qualifies as one illustration of my applied psychicality.

It continues to be essential that I study my (psychical) physiology in *as if* isolation from every other special interest of mine, but it cannot be studied *as if* separated from my mind itself for it consists only of my mind. Whether or not I keep myself conscious for this most consequential of all of my scientific factuality, is the issue of primary importance for observing the truth of my wholeness, and hence the requirement for my well-being.

My wholeness consists of my inneity, my inness, my innerness. Whatever is, *is* by virtue of its ipseity, its owndom. Man is capable of consciousness for this true view of the nature of all of his existence. Anew, all of my science of psychology can consist of nothing but idiopsychology. Awareness for this reality prevents my conceiving my wholeness as being like a Chinese nest of boxes, one within the other, each box containing nothing. My every meaning is constituted of *its* withinness. Instead of the illusion of interaction or interplay which depends upon plurality for its explanation, I make myself up of real self-continence which depends upon unity for its explanation. *To each only its own wholeness and innerness* is a biological law based upon repeatable psychological observation.

The term *understanding* is uniquely and only a psychological term, hence I need not expect any physiological or neurological understanding of any of my functioning, and may safely renounce such a chimera as being a brilliant fancy of mental genius. I do not object to trying out mixtures of interdisciplinary meanings but I can do so only at the risk of not focusing my attention on, and devotion to, directions of my living that I am competent to understand as repeatably workable for consciously fulfilling my wholeness-and-innerness nature.

My study of my mind, without acknowledging it to be mine, is not quantifiable. However, my study of my conscious mind is responsibly quantifiable. For example, I can measure my strength of mind specifically by how much of it I can claim as my own self identity.

Certainly, I honor all of the unconscious working of my mind, for example, my own psychically unacknowledged neurology, physiology, or oceanography because it is also out of such presently unacknowledgeable mental living that my further acknowledgeable mentality can grow itself. Each of my anatomical, physiological or biochemical terms has my mentality as its wholeness and innerness.

All that *can* be studied by me must be the application of my mind to whatever that study can mean to me. Hence I understand my physiology or biochemistry as psychic growth with which I can help myself to understand the comprehensiveness of my psyche, or mind, as such, and thereby prevent myself from repressing it. My concern as a physician for the painful but helpful signs and symptoms of repudiated psychic experience, accounts for my perseverance in working this mine.

Thus I grew greatly needed self understanding from the writings of my John Hughlings Jackson. Out of his brilliant doctrine of Concomitance leading him to treat his neurology as a physical science and his psychology as a science of the mind, I sensed more clearly than ever the necessity to reduce this bipartite account to a unity that cannot exclude the purely mental effort expended in each.

The tribute I pay to my usage of words necessary to compensate for dividing what cannot be divided, for excluding what cannot be excluded, is indescribably enormous. Much of my energy must be invested in trying to renounce each word in which I cannot see only my own mental meaning named. Such terms as placement, focus, locale, fixation, region, division, part, plural, and other names for fragmentation are particular sources of concern lest I mislead myself by them into ignoring my wholeness-innerness nature.

The conduct of my life, as the whole entity that it is, requires my reliance upon words naming concepts identical with, not alien to, my absolutely inviolable organic integrity as such. Thus I cannot say, "part of my body" or "member of society" without consciously heeding that "part" or "member" consists of nothing but my own whole individuality, idiomatically worded. My wholesome literary growth consists of having to invent a new way for me to talk or write to myself about myself, namely, my conscious idiolect.

In him who would evoke—create,
Contraries must meet and mate.

Herman Melville

PSYCHOPHYSIOLOGICAL INSIGHTS

The conscious end of the insightful psychologist or physiologist is to detect his (her) burgeoning unity, identity, wholeness, in every kind of experience. Thus he grows to recognize his every science as a unique view of systematic observations located necessarily in his own mind. He does not confound his methodized system or synthesis with alienation, any more than he confounds any other kind of individuation with alienated individuality. My physics or physiology can be no more or less than a cultivated linguistic mode of my mind, quite as my psychology or sociology or theology can be no more or less than a cultivated linguistic mode of my mind. The very law of *my* being must be my own origination.

Aristotle insightfully asserted that the indivisible wholeness of mind functions in each and every mental power. This means that no mental activity (such as sensation, perception, feeling, thinking) occurs separately, as might be implied in my scientific effort to discriminate or differentiate in order to abstract and analyze for concentrated study. I do well to practice heed for my constant need to cultivate my wholeness-consciousness. To assume that such difficultly attainable self-understanding will assert itself adequately as a matter of course is to substitute a policy of idly drifting, instead of ideally steering, for the conduct of my growing mind.

I must *grow* (in my mind and as my mind) whatever observation I make, regardless of whether I classify it as anatomical or astronomical. Every observation I grow reveals its wholeness in the necessity that I must seem to be able to remove it from its appearance of synthesis in order to abstract and analyze it as distinctly a separable entity. However, whether I heed the necessity or not, the known or sensed must be identical with the knower or senser.

Once I obscure the truth of my growing only my wholeness, in order to indulge the illusion of my divisibility, I may fall prey to the notion of my scientific work as being: the learning of more and more about less and less. Every insightful scientist is aware of the danger in imagined divisibility and the desirability in honored wholeness. In his *Religio Medici*, Sir Thomas Browne recorded, "The world that I regard is myself . . . Nature tells me, I am the image of God, as well as Scripture. He that understands not thus much hath not his introduction or first lesson, and is yet to begin the alphabet of man."

Again I note that by seeming to be able to entify my mentality into several existences such as willing, feeling, knowing, perceiving, sensing, and so on, I may feel tempted to regard such abstractions as so many separate, distinct, and independent existences. Nevertheless it is only my whole mind that wills, senses, understands, or whatever, merely by exerting itself in appropriate ways. With the conviction of the divine wholeness of man, stands the evidence of the existence of one Deity. As one varies in his conviction of his wholeness, so must he vary in his devotion to his godliness.

However, only I can call my attention to any of my

living and, until I can make myself aware of the fact that I am injuring myself by my heedlessness for my wholeness, I must continue on my seemingly divisive way of life, attributing my symptoms of rejected integrity to so-called impersonal agents. To be able to appreciate that I merely *grow* all of my behavior requires that I painstakingly work up my ability to attribute such amazing wonderfulness to my own living, hence personal responsibility for my true glory seldom reaches its full growth.

Nothing but my consciously hard living can provide the unique strength of mind called *hardihood*. My mind is my living of meaning called *experience*. Again, my mind is created by, and of, meaning. Hence mind is necessarily the single source of every such meaning as *matter, physics, object, externality*, or whatever experience I live that might appear to bypass mentality. For example, I may speak of my bodily activity as if it were not mental, physiological processes as if they were not psychical, sociological events as if they were not my mind's creations, and so on. *Pensive* primarily meant to weigh, *sapient* meant to taste, *sagacity* meant to scent.

The reality that all of my possible living experience, such as sensation, for instance, is produced by my growing it as integral to my whole self-development, must be of profound significance for resolving every possible problem of any moment in all of my life course.

Most desirable "quality education" is that which allows and favors the student's realization that all of his knowledge must be self-knowledge, and that all his study can accomplish is the development of his own mind according to the method of study he practices. The primary utility of every kind of knowledge is its necessary effect in developing the mind of the learner. To the extent that he is able to be aware of his own self-identity (his own subjectivity) in whatever subject

89

he learns, his scholarly pursuit enables him to work his head on rather than to seem to "work his head off" in the process. My science of physiology is most conducive to my advantage insofar as it remains conceivable by me as entirely subject of, and to, me. I am my only possible end to myself. *Idiopsychology, consciously or unconsciously, is my only possible psychology.*

Only through consciously *idiopsychical* research can the conviction of my wholeness be worked up and recognized. Anew, in my early mental development, I obscured the truth of my wholeness by disowning my own psychic experience that I could not acknowledge as being mine. I confounded my specific feeling of dislike with whatever I could not live comfortably, thus dissociating my mental wholeness into 1) all that I could acknowledge being and 2) all that I could not acknowledge being.

My child mind was not strong enough to observe that my dislike is all and only about itself. Similarly I would identify unpleasant taste, smell, or whatever sense with mentality other than itself thereby forfeiting my appreciation for my wholeness. Hatred is wholly hatred and cannot have any subject or object other than itself. But hatred (or any other painful feeling) unrecognizable as being wholly itself, accounts for all of my symptomatic living. "Know thyself" must refer to the *whole* self. All of my mental trouble is directly traceable to my not knowing my whole self. How horrible was the oracle to Oedipus,

"May'st thou ne'er know the truth of what thou art."

For my sensory growth to continue it is necessary that function of my sensory living continue, otherwise my mental balance becomes impaired by such diminution of biofeedback. Also atrophy of disuse can complicate

90

my effort to learn anew the functioning of the sensory feeling involved in such atrophy. With constant disuse of my skeletal structure beyond a week begins decalcification of bone and atrophic changes in my joints, tendons, and muscles, including generalized weakness. The point needs to be made and heeded that *with disuse of any of my functioning is the disuse of my mind, including my willpower, back of it.*

Sensation is emotion; perception is thought. However, each must be a special form of feeling to be alive. *Feasting* my eyes makes sense. The truth of my *growing* nothing but my own individuality may be viewed as the cardinal reality for creating whole-self understanding. It deserves to be interwoven with the functioning of my language as an indwelling reminder of my inviolable nature.

Once I grow capable of observing that my only possible self expression is reducible to being a growth of my individual mind, then each phantom problem created by my disregarding this vital necessity becomes recognizable as a symptom of my attempted psychic negation. For example, all that I blindly attributed to (*as if*) acquired knowledge becomes understandable as ignored intuition, or all that I blindly assigned to cerebral localization becomes explicable as already physiological mind that I have simply overlooked as such, or all that I blindly ascribed to so-called objectivity becomes evident as inhibited subjectivity.

Instead of my understanding psychic energy after the analogy of (seeming to be able to observe) so-called physical force, any and all of my so-called physical force can amount to nothing but what I mean (create mentally) and name "physical force." In other words, every meaning which I can live consciously or unconsciously is only and entirely a modification of my mind, and I must *grow* whatever modification that I can experience. All that my "experience" can mean is reducible to some form of my growing (living) myself.

All is native, instinctive, in *individuality*; nothing is accidental or adventitious. Thus my sensation must be merely my living of a specific function of my organic nature itself. My every sensation is a specific form of touching, and each touch is a form of my feeling only my own nature. All "touch" is limited to self touch. My olfactory patch, or taste bud, or pain corpuscle can feel only itself. Otherwise stated, *all* of my so-called externality is actually the subjective functioning of my own internality or native innerness. The whole doctrine of sensory *im*pression violates the truth of my absolute continence and inviolable individuality. All sensation must be self-sensation; all perception must be self-perception; all feeling must be self-feeling; all consciousness must be self-consciousness. Each of these conceptions must seem to be resisted by confirmed habit ("second nature") of judging and reasoning otherwise, unless each one is observable as all and only about itself. Although this kind of disciplined observation is still rare, it will flourish after reduction of resistance to merely the sober consideration of the possibility of conscious idiolect or conscious solipsism.

The revealing questioning to start about self-consciousness is: Must all self consciousness be acknowledged to be nothing but *self* consciousness? Must I actually be reflecting upon the fact that my mind is operating within itself, entirely and only, in order to be able to assert that I am then and thus being self-conscious? What difference must it make as far as my health or welfare is concerned whether or not I make such distinction? Can I take for granted with impunity that self-consciousness is a nomination that always means exactly what it is intended to mean, or must I take the trouble to generate in my mind my feeling of personal identity with every usage of that wording. What does the presence or absence of *my* deeply felt self-consciousness signify and, particularly, how does it affect my mental power?

All of my conscious activity must be only a specific modification of my mentality, its only possible subject being meaningful living of mine always specifically contributing to my appreciation for the wonderfulness of my life itself. My only choice is to affirm or repress this hard-earned self understanding that is required to imbue me with the will to take care of my magnificent being.

One day while I was holding forth on the vastness of the unconscious (my) Sigmund Freud remarked, "Unconscious? that is just a word." He had described well what he meant by his word "unconscious," namely, presenting mental activity whose existence is merely assumed. To be conscious for what *is* unconscious is obviously impossible. Although nothing can be known about the so-named unconscious, its actuality may be inferred by mental effects that can apparently be accounted for by making assumption of unknowable dynamic mental processes. Professor Freud cited the ordinary example of a slip of the tongue, how natural it is to assume the unobserved presence of the intention to make that slip. He distinguished two kinds of unconsciousnesses: one easily repossessed as consciousness and another difficultly, if at all, thus transformed. He also distinguished the meaning of consciousness as a quality of awareness and as a quantity of systematic mentation.

The indispensable benefits of so-named unconsciousness are too numerous to cite here. To illustrate, in order to be able to be solely conscious with regard to any one area of my living I must withdraw my consciousness from my living of all else which is nevertheless assumed to go on existing. All of the meaning of my life cannot be constantly conscious although its continuing existence must be assumed. All consciousness is self-consciousness, states Professor Freud. However, I can assume personal responsibility for only a given extent of my being at a time. I need my so-named unconsciousness to ease my mind constantly, and regularly, even to deep sleep. Whenever I need to repress (repudiate) any of my living, my so-named unconsciousness enables this freedom from unbearable personal responsibility. Hypothetically all of the living of my development from the moment of my conception is of the nature of such *as if* unconscious meaning.

Specific though any mentality of mine may be, by virtue of its becoming the specific focus of my study, nevertheless it must be a manifestation of the functioning of my whole mind. In other words, my mind is capable of *growing* innumerable alterations.

Special consideration of any such variation, even to giving it a special name, (such as imagination, perception, feeling, sensation, will, intelligence, consciousness, word, bone, sinew, nerve, and so on) does not give that creativeness any mindless independence or existence. Every meaning that I may habitually displace as if it might exist apart from my mind supports the illusion of the divisibility of the mind rather than the truth of mental wholeness. I can find nothing in my mind but myself. My so-called body expresses itself in endless variety, thus revealing its mental nature in every posture.

What I name my "body" consists of my self-grown mental abstraction only, but I may overlook this vital fact that "body" must be entirely a psychical development of my nature. The result, too notorious to be overlooked, is the conventional body-mind schism or duality, the symptom of repressed mentality. If I assume that I can safely divide my mind into autonomous powers independent of my wholeness, then it may seem logical that I can divide my body into somatic organs similarly "on their own."

What I name my "external world" consists of my mental abstractions only but I can regularly ignore the vital necessity that all of my meaning concerning my external world must exist entirely as psychical creation of my nature. The result, also too consequential to be safely ignored, is the conventional internal-external or subjective-objective duality, the symptom of repressed mentality. My fellowman at large believes in externality because he believes it can be immediately sensed and perceived by him, an illusion that even the name "externality" contradicts. If I assume that my observed externality can exist separate and distinct from my internality, then I can relieve myself from all psychical (including physiological) responsibility for my "external world" affairs and conditions. Unrecognized or recognized depression and anxiety then set up.

What I name my "Divinity" consists of my mental abstraction only, but I can deny the truth that it must exist entirely as meaningful power of my nature. The consequence, far too lifesaving to be neglected, is that it must exert its influence as if it could be a force alien to my percipient mind.

My so-called *Divinity* is vital to my understanding my wholeness, for example, in view of the fact that one of its basic synonyms is: perfection, or wholeness, or oneness itself. However my mastery of the meaning that I name Divinity involves mental strength to cope with the totality of my biologically adequate self-responsibility, and this degree of strength of mind is most difficultly attainable, hence rarely developed. If unattained I can relieve myself from responsibility for being all of myself merely by refusing to consider myself as perfect, as wholly one. *The symptom resulting from my negated divinity is no less than the creation of my problem of evil.*

My (self-consciously) limited fellowman regards extensive self-insightfulness as a kind of lunacy, equating his hold on what he calls his external world with his hold on sanity. He firmly believes that when "three men" (or more) look at any so-called object "they" are capable of seeing the same object. The fact is that any three-or-more-men are incapable of seeing anything whatsoever. Only individual man is capable of such self experience. My consciousness provides me only its own existence which I live. A serious linguist recognizes that language is made, as a rule, to serve the purpose of indulging ordinary illusions such as conversation, dialogue, communication.

Searching examination of my sensation, perception, consciousness, or whatever mental modality, reveals it as consisting entirely of my own being it. Certainly my experience has always taught me that there is existence other than myself and my *belief* in that existence is sure and certain. I have every *faith* in the existence of a universe that is *inclusive* of my own individuality. Although I base the conviction of my universal creed wholly upon my own persuasion traceable to my own personal experience, I am also perfectly at home in any and all of my other views about so-called external reality. In fact my present life-orientation is the outgrowth of my previous self-and-world views.

With my Emersonian fidelity I confess, "Belief consists in accepting the affirmations of the soul; unbelief, in denying them." I find whatever I believe in, to be the outgrowth of previous unbelief or wonder about it. My choice friend James S. Holden was fond of quoting Oliver Goldsmith,

> And still they gazed, and still the wonder grew
> That one small head could carry all he knew.

All of this fixing of responsibility for any living whatsoever *precisely and only in the liver* of whatsoever is lived, has direct bearing upon every effort to understand whatever I can mean by what I call my mind, the alleged creator of my every meaning. To illustrate, the scientific discipline I term *physiology* consists of nothing but my mental growth of mind, exactly as the discipline I term *psychology* consists of nothing but my mental growth of it.

However such insightful mind-orientation is based upon decisive evidence that can be found in one's own mind and confirmed in few treatises read only rarely. It is seldom, if at all, recorded in popular elementary works of science, philosophy, or the general textbooks of the school or college. I cannot find stated, except perfunctorily, in contemporary scientific literature the actual responsibility of an author for *all* of his own work. It is most helpful to consult the *original* author of any and every production, since that is its only possible faithful account.

If I feel my sensation, I can acknowledge that such sensing is my feeling. I cannot acknowledge that my sensing is my own in a mental condition of self-hypnosis. My mental condition of autohypnosis, consisting of denying that my living is my own, begins in my early life, when I first begin screening my experience. That which I am able to feel responsible for, is differentiated from my experience that I cannot feel responsible for. Concerning whatever I live, my choice is always that of acknowledged self-consciousness consisting of claiming it as my own personal identity, or of self-hypnosis wherin I disclaim it as my own personal identity. So-called mental trouble is traceable only to living of my own that I have had to disown, thereby making it unacknowledged self-conscious living of mine.

How does my creative imagination conceive this

almighty universe that I unquestionably believe exists? First of all I would expect it to consist of all of its internality, with nothing at all "external" about it. Having created all that I can imagine about it, it would have to be made in my own image and likeness. Thus I conceive it as consisting of nothing but individuality, wholeness, oneness, and whatever might correspond to my own identity, subjectivity, and responsibility for being all of myself. Certainly I believe it to be absolutely just, true, perfect, adorable, wonderful, infinite, eternal, self sufficient, self helpful, and self appreciative. *Nothing can happen or not happen in it unless sufficient truth is present or absent to make it happen or not happen.* I imagine it to be a growing, evolving, solipsistic allness, And thus I can go on constructing my cosmos as best I can according to my own acknowledgeable nature and needs for I, myself, am integral to that universe I am imagining to be. I necessarily postulate its existence, and accept it on faith.

So-called specific psychotherapy consists of gradually growing strength of personal responsibility to be able to acknowledge as mine, experience which formerly I could not acknowledge as mine, because I could not associate loving kindness with it. Such rejected self experience associated with my not wanting to live it consciously, because it threatens to make me feel too unhappy to want to go on living, is the source of all of my mental trouble.

As long as I can do so, I conduct my life in a way to spare my having any "show down" about the undeniable organic union of my living that I separate *as if* either the inferior or superior of my conscious image of my nature. Thus I may spend my life seeking easier ways to live it, becoming angry, anxious, jealous, depressed, or even apathetic, if I cannot escape such confrontation.

Sooner or later, if I live long enough, I may find myself overwhelmed by unavoidably conscious responsibility. Then I must either find a way to balance my increase of conscious responsibility, with the peace of mind that its associated appreciation for my self-greatness gives me, or succumb to being overwhelmed by my own unconscious excitation so that my conscious self identity no longer suffices me for the conduct of my life. Then, I can try other self help, such as seeming to locate my feeling of self identity only in some such alienated power figure of my life as God, Napoleon, or whoever.

Too much significance cannot be given to the grievous results of the ancient and modern onus of *as if* subject-object dualism. Nowhere is the *post hoc propter hoc* fallacy (what follows must be caused by whatever seemed to precede it) more prominently evident than in my persistent effort to prefer this schism in the face of the helpfulness of observable *self-grown* conscious mental integration. For relief from consequences I must assume for being my whole self, my so-called common-sense judgment supports any life orientation that promises me ease from my unbearably heavy duty to be the inviolable individual that I am.

Thus my obviously subjective sensation or perception may seem to be consequent on the presence of an object obviously external to me. Therefore it can appear only logical to assume the existence of some necessary connection between me and my externality. The idea that I can immediately sense or perceive any "external object" certainly does ease my mind of any responsibility whatsoever for my own necessity to be that *as if* "external object." Then by indulging my illusion that my conscious self-identity can sanely exclude any of my natural obligation that I associate with dislike, I can pursue my optative mood entirely out of keeping

99

with the truth of my difficultly livable and lovable wholeness. The hard-to-attain insight is that I generate my sensation and perception entirely as self-proliferations, quite as I produce only my identifiable body out of my food rather than organize resemblance to the meat and drink that I assimilate.

I believe that the degree and extent to which I can practice effective autosuggestion, disowning experience of mine as really being mine, is enormous. I ascribe the same kind and degree of autosuggestibility to my living of my fellowman. My experience steadily affirms this insightful observation, never opposes it, and certainly is readily repeatable. However, I am sure that my fellowman can only observe what he has grown capability for observing.

Observation is resisted, just as reasoning is resisted, as far as readiness to reclaim whatever mental experience that was once disclaimed is concerned. My number one public health problem goes on mostly even unsuspected by my public health official who has yet to grow the understanding that *whatever* he resists investigating is what specifically needs investigating. By "resist" I mean: overpowering unwillingness to consider any idea, or other mentality, soberly and steadily as being necessarily and helpfully itself. Resistance allows dislike to substitute for free interest.

In any respect, to attribute to my mental power a consciousness for any condition other than its own innovation, hypostatizes the existence of mentality that is observed as if not-mentality. Although it is possible for me to be able to imagine any so-called objective quality or quantity other than such possessed in my own mind, nevertheless no entity other than my own existence can ever occupy any place in my being. I can be absolutely *relative* only to my own absolute self.

All of my mental activity is immanent. None of my

power can be *transeunt*. My notion that I can somehow exist outside of myself, provides the easiest accounting for my being able to be aware of my own experience that I have rejected as being my own.

Mind defined as the absence of matter cannot be substantiated by mind because all that can possibly be *meant* by "matter" must be mental. Mind defined as the opposite of matter can be substantiated by mind because all that can be *meant* by "matter" then must be the (mental) negations of so-called mind.

Concerning the nature of non-mental matter no psychological datum is possible. All of my living of mathematics, for instance, consists of imagined abstractions of mine. My every sensory datum is my mental creation only. My *belief* in the existence of my whole universe, besides deriving from personal experience that necessitates my postulating such existence, is based on my belief that I am, myself, individually integral to this whole universality. I am a present instance of the nature of my universe, bridging the illusional chasm between my so-called internality meaning and my so-called externality meaning, each a term designating merely creativity of my mind. The apparent order of my so-called objective world issues from the evident order of my subjective self.

The necessity that all of my power be limited to the individuality of my own existence makes it impossible for me to act upon or react to another, indicating that all interaction and reaction in my world are illusional. Certainly this solipsistic necessity must be only resisted as absurd or nonsensical by everyone who has not discovered the inviolable wholeness-nature of his own individuality. Here and there my fellowman may know that he cannot really read any writing without taking the trouble to author it.

Give me a tiger and I will educate him.

Robert Owen

EXPERIENCE

SCIENTIFIC DISCOVERY

The cause of my scientific discovery is the same as the cause of my hand, foot, or head, namely, my *growing* each one. So-called causation is necessity, the creation of my wholeness, the sovereign domain of my nature. I grow my scientific knowledge in the same sense that I grow any regional meaning of my body. The growth of my judgment, argument, or language occurs quite as does the growth of my hair, nail, or tooth, namely, precisely when all of the preparation necessarily preceding it has already developed.

The realization that all of my behavior is caused, produced, by my *growing* it as integral to my living my whole individuality, is unique self-understanding. I refer otherwise to such brilliant stratagems as "cerebral localization of behavior," "the neurological foundation of mental activity," "psychological determinism," or whatever other reasoned design intending to account for the marvelous functioning of my life. Each, itself, becomes understandable as manifesting the growth of some of my necessary wholeness. *Healthy living must be based upon a biologically adequate estimate of the full meaning of being (living, creating, growing) an individual life.*

To Jean Martin Charcot (1825–1893), renowned Parisian physician of the mind and nervous system, is

105

credited a discovery of biological unification so amazing that, to this day, it remains largely unappreciated. Prior to his contribution (and even following it!) the "nerve specialist" found it helpful to divide his patients of troubled mind into two categories: *organic* and *functional*. The organic complaint was characterized by demonstrable structural changes in the central nervous system. The functional complaint was considered to be caused by thoughts and feelings independent of demonstrable brain-damage, hence classified as not organic. Charcot disturbed this easily peaceful dichotomy by demonstrating thoughts, ideas, conceptions to be organic also, teaching the evolutionary principle of the *organicity of the idea*.

Prior to this realistic appraisal of the structural nature of ideational force there was a strong tendency to depreciate the vital strength of an idea. Thus it was even thought that a person might just as well have one idea as another, considering the weightlessness of either one. Furthermore the meaning of imaginary became opposed to the meaning of real. Thus the real strength of mental power was relegated largely to what was called physically organic energy, and the understanding of the power of intact wholeness of individuality became correspondingly obscured.

My selfsame identity underlying *whatever* scientific studies I cultivate is to be found in the fact that each is wholly the outgrowth, the creation, of my mind's creativity. Hence it is that in order to understand my own nature it is necessary for me, first of all, to recognize that it (my own being) must be my only subject of study, whether I call it psychology, theology, physiology, astronomy, or whatever seemingly impersonal name.

The most frustrating consequence results from hot disputation as to whether behavior is psychical or physiological, organic or functional, real or imaginary, and so on. Therefore the ability to localize the cause of *all* living only in the whole individual responsible for that living can bring tranquillizing enlightenment to the sufficiently insightful one.

Sufficient consideration for the very concept of absolute wholeness must be resisted by me to the extent that I cannot honor the truth of my own absolute wholeness with my power of self-consciousness. Whatever living of mine I cannot fully acknowledge *is* mine, represents *unconscious* power of mine over which I do not exert conscious control. This unconscious power of mine functions forcefully but I have to regard its manifestations as beyond my voluntary control, describing them as "attacks," "spells," "fits," or some other foreign influences (such as "jabs of pain," "jerking sensations," "strange tensions," and so on).

Self-activity is the only possible activity. Strictly stated, all I can educate is myself about myself. My every sensation is, as philosopher-poet-physician John Keats observed his to be, nothing but an *intuition of my mind* forming the very current of my mental process. Keats being consciously a profound student of his own mind, could sing, "Knowledge enormous makes a God of me."

Only a philosopher-poet, that is, only one who can acknowledge that he can be a philosopher-poet capable of observing and cultivating his own glorious mentality, can know what it means "to be in continual burning of thought."[1] Understandably, the constant astonishment at the demonstration of my very own mental power readily becomes an unbearable responsibility so that I try to escape it by deluding myself that it is impersonal. As Keats divined,

> Beauty is truth, truth beauty—that is all
> Ye know on earth, and all ye need to know.

107

I do add "dazzling beauty" and "dazzling truth" in the effort to explain the rarity with which I can and do contemplate my almighty mental power as being wholly my own. I must perseveringly cultivate the mental endurance necessary to be able to bear such excitation with continent self-consciousness.[2]

Strong-minded Sir William Hamilton sanely asserted,

> A fact of consciousness, however accurately observed, however clearly described, and however great may be our confidence in the observer, is for us as zero, until we have observed and recognized it ourselves . . . instruction can do little more than point out the position in which the pupil ought to place himself, in order to verify, by his own experience, the facts which his instructor proposes to him as true.[3]

I am an organism in the sense of being one organization of my self-wholeness subject to the law of my nature, namely, inviolable individuality. So-called divisibility or plurality is necessarily foreign to my nature, except as representing my conscious effort to negate my necessary unity. Seeming coexistence of opposing forces in my being is really nothing but my efforts, conscious and unconscious, to restore the distinctive individuality (wholeness) of my being that has been rendered indistinct by my necessity to help myself by resorting to (illusional) dichotomies and analogous pluralities for the conduct of my life.

My psychology of *my* living scientifically establishes itself as my true preparative for leading an enlightened *personal*, including vocational, life. The source of all of my mental trouble is my resorting to my dichotomies, ever fictional, of self-and-not-self, God-and-devil, truth-and-falsehood, and so on.

The fateful illusion that one force or entity can have something to do with another underlies self-rejection. My daily vocabulary contains many life-discounting words. Merely by giving "a bad name" to any mental event, I repress it. "To repress" means to try to disown, as if the "repressed" were not life-worthy. Thus even persistent living of any of my experience with dislike augments the extent of my unconscious living that is responsible for symptom-formation.

As Sigmund Freud discovered, the symptom is a compromise representing both the effort of the repressed living to escape the repression and my conscious effort to maintain that repression. Hatred, fear, despair, or any other unpleasing feeling, is *all and only about itself.* Seldom is this truth consciously effective so that the unpleasant feeling can operate as if it could apply to living other than itself, thereby favoring the functioning of repression.

For example, I hurt myself and feel pain as with a toothache. I can take a "pain-killer" pill and forget all about taking care of my tooth. Or, my living of my friend is associated with anger. I can refuse to pay further attention to my living of my friend, so that I avoid my anger, but I forget all about taking care of my living of my friend, and so on.

Conscious idiopsychology is preeminently my science of my mind. Conscious idiolect is my mental operation of words denoting a reduction of seeming manyness of generality to the real oneness of my uniquely particular being. Conscious self-development is to the freely self-acknowledgeable individual what the parent is to the child. Fortunately, benevolent skepticism preserves me from overwhelming my mind with whatever consideration in which I am unable to recognize my own personal identity.

My sense of my personal identity is my only conscious basis for taking any care of myself. The process of augmenting my conscious self-identity is the procedure conducive to the end which I propose as being life fulfilling, namely, my growing appreciation for the wonder of my self-wholeness. I can be conscious only when I am living some activity of my self. I am fully satisfied to be able to attest my own existence only. I cannot be active where I am not present. My every idea follows from the sole necessity of the nature of my mind. There can be nothing in my life which is not created by me.

The elements of my existence constitute my self-identity. The conscious elements of my existence constitute my conscious self-identity. Becoming aware of my self-consciousness is the process of discovering and acknowledging being any or all of my own living. Henrik Ibsen (1828–1906) thus condensed his sanity: Love of self is the fundamental principle of all activity. Self comprises all there is to itself.

As a scientist I recognize the necessity for exertion. It is estimated also that the *Bhagavad-gītā* has become the most important Hindu Scripture on account of its due emphasis on work. Becoming aware of the individuality in the real, as well as of the reality in the individual, involves the hardest working of consciousness revealing its meaning for subjectivity. The self consciousness-in-experienced mind is unready for living its ideality, yes, its spirituality, except repressed as bugbearish.

I being a conceived and born innovator, all of my life may be accurately described as momentous self-discovery. Doing the hard labor required to make that self-discovery into *conscious* self-discovery is my number one health issue. A description of how to achieve this lifeworthy attainment is not difficult. To illustrate, in order to grow scientific "discoveries," one after another, all I need do is to create an *as if* environment including scientific instrumentation (e.g., a laboratory), then imagine a scientific goal that I might devote my attention to, and interest in. The rest need consist merely

110

of my living heedfully the observations and conceptions that naturally occur in the course of my living my mind thus modified.

Self-conscious Ralph Waldo Emerson divined: "Souls are not saved in bundles," "The true meaning of *spiritual is real*," "An anatomical observer remarks, that the sympathies of the chest, abdomen, and pelvis, tell at last on the face, and on all its features," and "The feat of the imagination is in showing the convertibility of every thing into every other thing." He also said that Science is a search after identity. I would add only: All identity is self identity. As an expert psychologist I may not hesitate to call my fellow expert who disagrees with my present psychology "a madman," unless I can become sufficiently self-conscious in time to prevent such implied self-disrespect.

Conscious unity of my self as an integral being permits responsible mobilization of my will so that I can apply myself as I please most forcefully, energetically, enthusiastically. I cannot but sense the major moral in securing optimal manpower from thus heeding my homogeneous wholeness.

That profound moral decrees: I can and must honor the whole truth in and of whatever I live, exactly to the extent that I can and must honor the truth of my own wholeness. But habit makes it easy for me temporarily to lose my sense of my true wholeness in my fixed fascination for merely some of it. My ideal morality is simply natural, namely, my obligation to strive to help my acknowledgeable self as best I can to appreciate the greatness of *all* of my life, pending my growing greater skill for fulfilling that most worthy achievement. This scientific insight, namely, that I must study hard to know *how* to live my life wisely, confers a certain feeling of wholeness either akin to, or identical with, holiness. My present opinion of myself is an acknowl-

111

edgment of the character of my present morality.

Thus Professor Borden P. Bowne sounded the keynote of a policy for enjoying the freedom of his soul: "The undivineness of the natural and the unnaturalness of the divine is the great heresy of popular thought."[4] Accurate and easily repeatable investigation discloses that the truth is always on the side of whatever happens, before it can and must occur. This scientific discovery means no less than that: in divine perfection we live and move and have our being (the so-called doctrine of divine immanence). There can be no division of labor or authority or responsibility in indivisible man. He is his own organic *all* that he lives. It is my indulging my illusion that the indivisible is divisible that creates my (illusional) problem of evil or error or undesirability of any description. A divine Savior dying on the cross to save mankind, is a beautiful treatment of the suffering I, including my wayward fellowman, must endure until I can create my opportunity to cultivate my own loving whole-self understanding.

Much of what may seem merely repetitious in these pages I intentionally retain for its calisthenic value as my strengthening mental exercise.

112

The motive of science was the extension of man, on all sides, into Nature, till his hands should touch the stars, his eyes see through the earth, his ears understand the language of beast and bird, and the sense of the wind; and, through his sympathy, heaven and earth should talk with him. But that is not our science.

Ralph Waldo Emerson

SUMMARY

Physiological research, like all anatomy, biochemistry, sociology, astronomy, or any other kind of investigation, has its entire reality in the mind of the individual investigator. Every research manual, book, library, or laboratory is inert until enlivened by the specific mind of the scientist exploring and exploiting it. My given science is my systematically structured and arranged language of self-knowledge occurring in my mind only. My life can sense or observe nothing but the life of itself.

With all due honor to Claude Bernard's greatness as a physiologist, I find it impossible to work in the laboratory like an animal, diligently but without a mind distracting me from immediate study of the presenting facts. Rather, I can conduct my laboratory research as a god creating my every observation, being the unmoved mover of all of my experience, finding my image and likeness in my every creation.

Whence comes my insistent demand that my mind be what it is not and not be what it is? Why must I expect to be able to get at something or somebody other than myself, and the converse? How can I explain my need to get out of myself, or to claim I am what I am not?

Not merely in speaking, but as certainly in thinking,

my tendency to conform to the grammar of the vernacular may be steadily indulged, mostly without my noticing such unwitting disregard for consequence. Assuming responsible attention for the vast extent to which my mentality consists of language is a heavy burden that I put off as often as I can. My need to, so to speak, remind myself of myself is unremitting but my recognition of that need is all too seldom aroused.

Thus, I may go on using my mind automatically, as it were, without much awakening to the whole-making truth that it *is* all mind. Indeed I may thereby dismiss many a life-activity of mind with the belittling notion, "It is merely verbal." Such attempt at heedless conversion of my life-experience into a kind of dead language prevents my duly appreciating my continuous life in my living.

John Locke, M.D. (1632–1704) may be referred to as the father of present-day psychology in that he founded his understanding of mind upon its innate sensibility, including emotionality and meaning. What cannot be reduced to such excitability is not mental. It is the absence of this organic irritability that is called death. On the other hand, nothing can be responsible for any meaning but mind itself. I can go straight in correcting stray behavior only by feeling my way.

Whether I use my mind to arrange its meaningfulness into a certain circumscribed order that I name physiology, spirituality, minerology, or extrasensory perception, the only factuality about any of it is that it is a more-or-less orderly series of sensation or feeling identifiable only as mind. By definition there can be no individual experience save that created by the given individual self. Unless I can view my experience where all of it must occur, namely, in my own living, I shall be in danger of insightlessly expecting my biological resources to become what they are not or to not be what they are.

The best I can compose on the *Psychic Nature of Physiology* must be the product of the experience I have lived in understanding the artificial nature of every appearance of division of individuality. I have grown strength in acknowledging the quality of inviolable unity or wholeness in individuality, including the necessity that such strength of acknowledgment must be grown, if it is to exist. And growth is naturally a gradual process obeying only its own law; it is characteristic of life only. The seeming divisibility of indivisible individuality must be merely an appearance operating under its own law of mechanics. All of the illusion of growth created by astounding mechanism, as in the computer, is traceable to the ingenious inventiveness of its originator. A machine appears to consist of divisible wholeness, although it is made in its creator's image by his using his identity for his creativeness.

My habit of mind of keeping the meaning of each scientific discipline in a watertight compartment, so to speak, is analogous to keeping each one *as if* external to every other one. Therefore, it is essential for my ability to appreciate my inviolable wholeness, that I do not allow the appearance of difference in my mind to obscure the truth of its absolute homogeneity.

My persistent yearning to get physiology and psychology "together," or to unite any other disciplines, is thoroughly understandable as a "return of the repressed," as a necessary consequence of my need to reject the abiding truth that "they" are one already in the oneness of my unique mentality. All of my understanding is the product of my self-understanding. I arrive *instinctively* at all of my sensation, perception, feeling, knowledge, or whatever experience I live. I, only, originate any and all of my meaning.

To be sure, physiological meaning is ever a discrete instance of psychic meaning. Hence I may wish to see only physiological meaning in that psychic meaning, after I have seemed to be able to rule its psychicality out. By seeming to rule it out I create the appearance of non-psychic meaning.

In my wish to undo such contrariness may be discovered a variation of an age-old longing to assert that my mind can have access to so-called externality meaning. Actually the meaning named objectivity, defined as perceived external datum, is subsumed by the meaning of subjective observation.

In all of my sensing, thinking, feeling, doing, such as this present writing about my lifesaving selfconsciousness, I certainly enjoy the feeling of discovering my whole unity here-and-there where before I was not aware of its existence. However, beyond, anterior, and prerequisite to that enlivening satisfaction is my pleasure in my way of life itself; in the course that my living takes or wills consentaneously; in the conduct of my nature, as such, meaning only that I am living solely my creativity to make my self be; in the absolute certitude of my untroubled perfect right to prevail quite as I am.

This just appreciation for my wonderful being, *ipse*, is my unfailing source of peace and quiet. If my troubled colleague appears to say to me, "Show me the afferent pathway to peace and quiet, to love, hate, guilt, jealousy or fear," I then ask this question of myself. Then I find myself able to realize that my colleague's question really reflects his need to translate his language of neuroanatomy into his language of self-conscious selfness by using such word meanings as understanding, attraction, repulsion, identity, wholeness, and the like. My neuroanatomy as well as my every so-called physical science speaks the language of behaviorism ("objectivity"); my consciously self-contained scientific language acknowledges its basis in my acknowledged self-conscious solipsism consisting of subjectivity, which equates with individualism.

For understanding the true nature of my being, by beginning merely with so-called physical evidence I make myself start by building my foundation upon illusions of externality and objectivity (through ignoring my necessities of internality and subjectivity). Each of my so-called physical conditions certainly exists (in my mind, only, to be sure) or I would not be able to consider it at all. The essence of its presence in my mind, alone, is sufficient to require my honoring each physicality-meaning as psychic reality consistent with my own intrinsic nature. So-called fact or so-called fiction, so-called reality or so-called imagination, subjectivity or objectivity, private or public, belief or doubt, spirit or substance, -it is all one to me in the living sense that each is a word or name for a meaning existing in my mind only.

The enormous price that I must pay for indulging my devotion to *as if* physicism is that of disregarding what is most precious of all of my life's meaning, namely, the meaning of my life itself. It is my life only that can ever be found in my mind and *the prosperity of my health depends specifically upon the extent to which I abide by that realization knowingly.* Growing from ignoring this law of being (namely, I am all and only that I am) to revering it for my definition of my soul, relieves me of whatever sign or symptom of self-belittlement (complaint) I may be presently indulging. Acknowledging completely that only I can grow any knowledge of myself, enables me to realize the absolute futility in trying to know anything about anyone else.

Meister Eckhart recorded his soul-wisdom thus, "Everything that is, has the fact of its being through being and from being." Unless I can succeed in creating

119

understanding for this kind of conscious wholeness, by and of myself, I cannot attribute it to *any* of my living, to my fellowman or to my God.

All I can ever know about anyone or anything but my own self is that it (whoever or whatever it may be) is entirely and only *its* inviolably powerful and self-sentient individuality.

My belief in my materiality (other than that of the imponderable substance of my own living) is the source of my attributing materiality to any and all of my nature that I call my external world. The indescribable extent to which education to so-called materiality prevails in my world fully accounts for the alarming signs and symptoms of unconscious use of almighty being. It is natural to cry out against such suffering; it seems unnatural to recognize the glorious goodness in it. The price of irresponsible behavior exactly equals the cost of self-unconsciousness. Each feeling of grief bespeaking sense of irreparable loss is successfully treated by the recovery of the sense of my unimpairable wholeness.

NOTES

Chapter 1

1. See my chapter, "Idiolect," in *Communication of Scientific Information*, ed. Stacey B. Day (Switzerland: S. Karger AG, Basel, 1975), pp. 12-27, in which I present the understandable helpfulness of conscious idiolect for what I call "carcinomentation." Locked in my materialistic conception of my sense world, my free spirit can seem frozen in its own linguistic structuralizations.

It is not my object to author a definitive account even of my singular concern, namely, foundational knowing of what I am writing about. To illustrate, the very fact that I think that I know something about somebody else or something else (other than myself) always turns out to be proof certain that such knowing is information about myself that I refuse to acknowledge.

The whole aim of this book is essentially the same as that of my preceding psychological works, specifically, to record how I help myself to understand the only meaning that can become understandable for me, namely: knowledge of the nature of my own living. For this purpose I willingly renounce any and all other aim or claim. My conscious idiolect, that is, all of my acknowledgeable wordage, does not require that I construct a specific language other than the continuing outgrowth

121

of the linguistic development that has always been mine. It does require, however, my arduously constructing specific personal responsibility for my being the one and only creator of my every word-creation.

The formulation of this insightful verbal orientation, enabling me to recognize my vocabulary as a binding force of my being, provides the guiding principle of the present writing. It is indispensable to my clarifying for myself the deep meaning that all of my so-called language is all and only itself and about itself, the language of language; that all of my so-called science is all and only itself and about itself, the science of science; that all of my so-called philosophy is all and only itself and about itself, the philosophy of philosophy; and so on. My construction of any language system (e.g., physiology or physics) presupposes my own mind as its only creator. Whatever is, is all and only of and about itself. All of my science is predicated upon this metaphysical necessity: every concept must do only what it is. Its functioning is its true meaning. I must either honor its visceral nature or suffer the painful consequence of its inhibited vitality.

I can use every faculty of my sensorium and motorium without acknowledging full responsibility for that functioning. I can even practice great precision in defining my terms or concepts on any subject, but if I omit (e.g., as "taken-for-granted" or "going-without-saying") my full acknowledgement to myself (including my every fellowman) that I am the only one responsible for creating those terms and their definitions, then I am using my unique psycholinguistic power dangerously, thus weakening rather than strengthening my appreciation for my life itself. As a matter of fact, I find such unconscious behaviorism ("objectivity") firmly established in the mind of my contemporary fellowman, with only rare exception. *I trace all mental trouble, including criminality and war, of my world to weakened appreciation for the marvelousness of individual life!*

2. "The New Style of Science," *Yale Alumni Magazine*, Feb., 1963.

3. See my *Psychology of Language* (Detroit: Center for Health Education, 1971). Also, "Idiolect," *Communication of Scientific Information* (Switzerland: S. Karger AG, Basel, 1975), pp. 12-27. In her life-revealing diary author Doris Schwerin beautifully describes her own convincing account of the origin of cancer (*Diary of a Pigeon Watcher*, New York: William Morrow and Co., 1976),

> When all is well, balance maintained, a sensible peace, the norm, I would bet my bottom dollar there is no cancer. When there is grief, denial, guilt, depression, I bet whosoever is *susceptible* will sequester cancerous cells, when one's life is most vulnerable.

Understandably, it seems more natural for me as an objective scientist to study my objectified cancerous cells, rather than to overwhelm my mind with my unconscious subjectivity, particularly since my creation of my objectivity is for the express purpose of avoiding confrontation with my subjectivity!

Chapter 2

1. New York: Funk and Wagnals Co., 1908.

2. *Studies in the Thought World or Practical Mind Art* (Boston: Lee and Shepard, 1903), pp. 49-50, 57, 85, 92, 99, 124, 127, 130, 143, 151, 166, 167-8, 178, 179, 243, 303.

Chapter 3

1. See my *Illness or Allness* (Detroit: Wayne State University Press, 1965), pp. 21–43.

Whatever my interest, it must derive from the metaphysical foundation I term *my mind*. The principle of sufficient reason exalted by my Leibnitz, Spinoza, and other conscious self-realists is difficult to uphold and easy to dilute to so-called truism. My growth of my concept that growing myself is my only possible kind of behavior, is one of my few ideas that seems to be historically less than twenty-five hundred years old. As Alfred North Whitehead observes in his *Science and the Modern World* (1925):

> My explanation is that the faith in the possibility of modern scientific theory, is an unconscious derivative from medieval theology.

Upon reading his Anaxagoras, Socrates grew for himself the idea that mind is the ultimate "disposer and cause of all" existence, for the best. He said to himself that "if anyone desired to find out the cause of the generation or destruction or existence of anything, he must find out what state of being or suffering or doing was best for that thing, and therefore a man had only to consider the best for himself and others, and then he would also know the worse, for that the same science comprised both." Plato's *Phaedo* (399 B.C.). Socrates, too, then noticed with grievous disappointment that his philosopher did not proceed systematically to refer to mentality as the source of the best accounting for each and all. Rather he appeared to neglect this one and only true explanation for all life-understanding by trying to explain the disposition of the limbs and other regions of the body by resorting to contraction and relaxation of muscles, hardness of ligaments and bones, and so on. Thus he was "forgetting to mention the true cause," *mind* which alone can "dispose all for the best and put each particular in the best place."

2. Rudolph Ballentine, Jr., Swami Rama, and Swami Ajaya (Allan Weinstock), *Yoga and Psychotherapy: the Evolution of Consciousness* (Glenview, Ill.: Himalayan Institute, 1976). *Yoga and Psychotherapy* is a timely, notable, and noteworthy attempt to bring relief to the Western scientist's tendency towards depersonalization, by presenting self-findings of the systematically disciplined student of Yoga.

3. My Wayne State University colleague, Professor Sheldon J. Lachman, has published a concise survey of significant contributions in the history of psychophysiology, *History and Methods of Physiological Psychology* (Detroit: Hamilton Press, 1963).

4. *Bulletin of the History of Medicine* (Baltimore, Md.: Johns Hopkins University Press, 1975), 49:4, p. 473.

5. See Margaret Laird's *Christian Science Re-Explored, A Challenge to Original Thinking.* Introduction by John M. Dorsey, M.D. (New York: William Frederick Press, 1965).

6. "Review of Neurophysiology Discovers the Mind," *Feelings and Emotions: The Loyola Symposium,* ed. Magda B. Arnold, (New York: Academic, 1970), pp. xvi, 339. Reprinted *Contemporary Psychology,* June 1971, vol. xvi, no. 6.

7. Oxford University Press, 1941.

Chapter 5

1. See my *Psychology of Language* (Detroit: Center for Health Education, 1971).

Chapter 8

1. Letter of Keats to Leigh Hunt, May 10, 1817.

2. See my *Psychology of Emotion* (Detroit: Center for Health Education, 1971), chapter 3.

3. *Lectures on Metaphysics* (New York: Sheldon and Co., 1858), pp. 1, 11.

4. Bowne, Borden P., *The Immanence of God* (Cambridge, Mass.: Riverside Press, 1905).

Any and all of my world that seems to be other-than-myself can be no more than meaning that I find in my own mind and predicate of my imagined, so-called "otherness." I accomplish this feat of imagination through my mental process of *as if* objectification of that much of my own self-identity. Only I can be the subject-object of whatever I live. My every experience is a synonym for my growing specific living of my individuality. Quite as Baruch or Benedict de Spinoza proposed of his Divinity: Besides Itself, no substance can be nor can be conceived.

By extending my I-feeling, that is, my acknowledgeable self-consciousness, I can unite my experience into an appreciable wholeness-entity thereby making conscious the real organic integrity of my nature. This uniquely self-teachable process of *conscious* psychogenesis, most trying because it entails the necessity of my heedfully augmenting my full responsibility for my behavior, enables my realizing my greatest ideal of self-fulfillment.

NAME INDEX

Abraham, Karl, 39
Adrian, Edgar Douglas, 36
Africa, 71
Ajaya, Swami (Allan Weinstock), 125
Alcott, William Andrus, 31
Alexander, Franz, 37
Anaxagoras, 27, 124
Aquinas, Thomas, 38
Aristotle, 28, 37, 54, 68, 80, 87
Arnold, Magda B., 125
Atlantic Monthly, 41
Augustine, Saint, 38, 48, 66
Ax, Albert F., 36–37

Bacon, Francis, 29, 53
Bacon, Roger, 29
Baillie, John, iii
Ballentine, Rudolph, Jr., 125
Bando, H. Walter, ix
Barrett, Albert M., 52
Basel, Switzerland, 121, 123
Bell, Charles, 31, 51
Benedek, Therese, 37
Ben-Gurion, David, 39
Bentham, Jeremy, 30
Berger, Hans, 36
Berkeley, George, 38
Bernard, Claude, viii, xi, 32, 115
Bhagavad-gītā, 110
Blake, Mary Elizabeth McGrath, 37
Bowne, Borden P., 112, 126
Boyd, Julian P., 38
Brazier, Mary A. B., 39

Breal, Michel Jules Alfred, 38
Brennan, Thomas P., 35
Brentano, Franz, 38
Broca, Paul, 33
Brown-Sequard, 51
Browne, Sir Thomas, 71, 88
Bulletin of the History of Medicine, 125
Bunyan, John, 29
Burns, Robert, 15
Burton, Robert, 29

Cabanis, Pierre Jean Georges, 30
Cajal, R., 51
Calvinist, 49
Cannon, Walter B., 35, 40
Cassirer, Ernst, 39
Center for Health Education, ix, 126
Charcot, Jean Martin, 38, 105–106
Chesterton, Gilbert Keith, 52
Child, Charles Manning, 34
Childs, Marquis William, 39
Chinese, 82
Coleridge, Samuel T., 37, 47
Collier, Arthur, 38
Condillac, Abbé Étienne Bonnot de, 30
Conway, Moncure Daniel, 38
Crim, Mrs. William D., ix

Darwin, Charles Robert, 32
Day, Stacey B., 121
Demosthenes, 5
Descartes, René, 29

SUBJECT INDEX

132

133

Knowledge, vii, 1, 46, 89, 119

Language, xii, 1, 49, 67, 73–86, 91, 113, 118, 121, 122
Law, 5, 7, 13, 25, 82, 87, 119
Learning, 35, 43, 53, 79, 81, 88, 89, 107, 126
Life, 1, 3, 18, 22, 27–40, 43, 45, 47, 55, 65, 68, 75, 77, 80, 115, 119, 123, 124
Lifesaving, 8, 9, 10, 19, 43, 48, 53, 59, 69
Localization, 7, 22, 27–40, 46, 50, 56, 80, 84, 87, 97, 99, 107
Loss, 120
Love, 98, 100, 110, 112. Also see Emotion, Self, and Wholeness

Magic, 20, 81
Magnanimity, 63
Man, 17, 113
Mastery, see Renunciation
Materialism, 4, 32, 50, 101, 121
Materiality, 27–40, 78, 120
Meaning, 1, 2, 6, 8, 27–40, 44, 49, 56, 57, 74, 76, 97, 117, 118, 121
Mental hygiene, 34, 77
Mentality, vii, 1, 3, 21, 22, 77–78, 122, 124
Metaphysics, 23
Mind, 1, 2, 3, 5, 21, 27–40, 44, 46, 49, 51, 55, 65-69, 76, 87, 97, 101, 109, 115, 119, 122, 124
Mirror-image, 9, 10
Morality, 29, 31, 37, 47, 48, 49, 111
Morbus medicorum, 9
Motorium, 122
My, 7, 48, 49, 55, 56, 73, 108
Mystic, 28

Name, 6, 7, 106, 108
Natural, 22
Nature, 12, 13, 17, 18, 20, 30, 34, 48, 49, 54, 83, 119, 121, 126
Necessity, 6, 12, 17, 46, 47, 65, 67, 73, 82, 101, 119, 122, 127
Negation, 11, 22, 69, 91, 96, 108
Neurology, 27–40, 51, 80, 83, 105

Objective, 7, 22

Objectivity, 2, 3, 7, 8, 9, 29, 35, 45, 46, 47, 68, 118, 122, 123
Obligation, 49
Observation, vii, 5, 43, 100
Oneness, see Whole
Ontogeny, 9
Opposites, 68, 80, 85, 108, 118
Order, 21, 101
Organ, 56, 95
Organic, 2, 50, 51, 55-57, 65, 78, 84, 106, 116, 126
Organism, 2, 51, 108
Originality, 66, 97, 117
Otherness, 4, 6, 84, 126
Overwhelm, 4, 19, 51, 75, 99, 109, 123

Pain, 11, 12, 22, 45, 55, 65, 66, 108
Paranoia, see Reasoning
Pathology, viii, 4, 8, 9, 17, 18, 19, 69, 74, 75, 90, 91, 120, 122, 123
Peace, 20
Pejoration, see Faultfinding
Perception, 75
Perfection, see Wholeness
Personal, 57, also see Identity
Phantom problem, 8, 54
Philosophy, 4
Phrenology, 30
Physics, 4
Physical, 6, 81, 119, 125
Physicism, 119
Physiology, viii, 2, 12, 17, 18, 22, 31, 43–84
Pleasure, 11, 45, 53, 55, 66
Plurality, vii, 8, 45
Practice, see Exercise
Psychiatry, x, 50, 52
Psychic, 20, 23, 119
Psychicality, 1, 3, 6, 7, 10, 18, 34, 43–84, 118
Psychicalize, 45
Psychoanalysis, 18, 34, 64
Psychobiology, 34
Psychogenesis, 6, 48, 54, 57-59, 64, 66, 126-127
Psychologist, 18, 51
Psychology, 1, 21, 77, 97
Psychophysiology, 36, 37
Psychosomatic medicine, 37
Purpose, 4

Quantifiability, 83
Question, 10, 11, 44, 80, 92

Reader, 9, 43, 65, 79
Reality, 3, 7, 21, 22, 45, 48, 55, 67, 77,
  89, 106, 110, 115, 119
Reasoning, 7, 68, 116
Relationship, 6, 7, 8, 45, 51, 67
Relativity, 35, 39, 100
Renunciation, 17, 19, 49, 53, 79, 83, 121
Representation, 51
Repression, 1, 4, 19, 50, 65, 67, 80, 83,
  94, 98, 108
Research, 5, 7, 29, 56, 64, 79, 81, 90,
  100, 115
Resistance, 15, 17, 18, 19, 20, 31, 44,
  50, 66, 67, 74, 80, 81, 100,107
Responsibility, 8, 12, 28, 29, 33, 46, 50,
  54, 94, 96, 99, 107, 116, 120, 121, 127
Reverence, 27, 119

Science, 7, 29, 30, 32, 33, 53, 65, 76,
  81, 97, 105-112, 113, 118, 121, 122,
  124
Scientist, 1, 7, 22, 29, 73, 81, 115, 125
Self, viii, 2, 3, 6, 18, 19, 20, 22, 27-40,
  48, 49, 65, 76, 79, 107, 120, 126
Self-analysis, 6
Sensation, 11-30, 50, 55, 75, 90, 92, 97,
  116
Sense, 46, 88, 118, 120, 121
Sensorium, 122
Separation, 3
Sex, 10
Sleep, 2, 9, 43, 64, 74, 94
Society, see Otherness
Solipsism, 2, 3, 37, 45, 78, 92, 101, 119
Soul, 12, 21, 119
Spirit, 41, 46, 111
Subject, 3, 12, 68
Subjectivity, 3, 8, 21, 22, 51, 55, 89, 123
Suffering, 12, 120
Sufficient truth, principle of, 98, 124
Survival, 32
Symptom, 4, 8, 19, 20, 22, 45, 46, 48,
  49, 50, 56, 69, 74, 80, 94, 95, 108, 119,
  120

Synaesthesia, 75
Synthesis, 64, 88

Transeunt, 100
Treatment, also see Therapy and
  Growth, 12
Tropism, 34
Truth, 20, 45, 50, 55, 63, 69, 74, 107
  111

Unconscious, 4, 46, 74, 93, 94, 107, 120
Unconsciousness, 2, 18, 22, 27-40, 51,
  57
Understanding, vii, xi, 5, 6, 7, 27, 52,
  63, 80, 82, 118, 120, 121, 124
Unity, see Wholeness
Universality, vii, 21
Universe, 3, 28, 73, 96, 98, 101
Unpleasure, 18
Unselfish, 8

Value, 65-66, 112
Vocabulary, 3, 5, 46, 51, 74, 108, 121
Volition, 12, 31, 43, 66, 68, 79, 91, 93,
  111

War, 123
Wealth, 21
Whole, 1, 18
Wholeness, vii, 1, 2, 3, 5, 6, 7, 10, 11,
  12, 19, 20, 22, 27-40, 43, 45, 46, 48-59,
  63-69, 76, 78, 82, 87, 89, 94, 101, 110,
  111, 117, 120
Will, 79, 118, also see Volition and
  Growth
Wisdom, 21, 28, 65, 119
Wish, 5, 29, also see Growth
Within, 3, 6, 21, 27-40, 44, 71, 75, 82,
  98
Wonder, 48
Word, 3, 5, 45, 46, 76, 83, 93, 109, 121
Work, 45, 49, 50, 53, 65, 90, 110, 127,
  also see Discipline

Yoga, 27, 125

Developing Effective Internal Evaluation

Arnold J. Love, *Editor*

NEW DIRECTIONS FOR PROGRAM EVALUATION
A Publication of the Evaluation Research Society
ERNEST R. HOUSE, RONALD J. WOOLDRIDGE, *Editors-in-Chief*

Number 20, December 1983

Paperback sourcebooks in
The Jossey-Bass Higher Education and
Social and Behavioral Sciences Series

Jossey-Bass Inc., Publishers
San Francisco • Washington • London

Arnold J. Love (Ed.).
Developing Effective Internal Evaluation.
New Directions for Program Evaluation, no. 20.
San Francisco: Jossey-Bass, 1983.

New Directions for Program Evaluation Series
A Publication of the Evaluation Research Society
Ernest R. House, Ronald J. Wooldridge, *Editors-in-Chief*

New Directions for Program Evaluation (publication number
USPS 449-050) is published quarterly by Jossey-Bass Inc.,
Publishers, and is sponsored by the Evaluation Research Society.
Second-class postage rates paid at San Francisco, California,
and at additional mailing offices.

Correspondence:
Subscriptions, single-issue orders, change of address notices, undelivered
copies, and other correspondence should be sent to Subscriptions,
Jossey-Bass Inc., Publishers, 433 California Street, San Francisco
California 94104.

Editorial correspondence should be sent to the Editors-in-Chief,
Ronald Wooldridge, Bureau of Forecasting and Modeling,
44 Holland Ave., N.Y. 12229, or Ernest House, CIRCE-270,
Education Building, University of Illinois, Champaign, Ill. 61820.

Library of Congress Catalogue Card Number LC 82-84201

International Standard Serial Number ISSN 0164-7989

International Standard Book Number ISBN 87589-968-4

Cover art by Willi Baum

Manufactured in the United States of America

Ordering Information

The paperback sourcebooks listed below are published quarterly and can be ordered either by subscription or single-copy.

Subscriptions cost $35.00 per year for institutions, agencies, and libraries. Individuals can subscribe at the special rate of $25.00 per year *if payment is by personal check.* (Note that the full rate of $35.00 applies if payment is by institutional check, even if the subscription is designated for an individual.) Standing orders are accepted. Subscriptions normally begin with the first of the four sourcebooks in the current publication year of the series. When ordering, please indicate if you prefer your subscription to begin with the first issue of the *coming* year.

Single copies are available at $8.95 when payment accompanies order, and *all single-copy orders under $25.00 must include payment.* (California, New Jersey, New York, and Washington, D.C., residents please include appropriate sales tax.) For billed orders, cost per copy is $8.95 plus postage and handling. (Prices subject to change without notice.)

Bulk orders (ten or more copies) of any individual sourcebook are available at the following discounted prices: 10–49 copies, $8.05 each; 50–100 copies, $7.15 each; over 100 copies, *inquire.* Sales tax and postage and handling charges apply as for single copy orders.

To ensure correct and prompt delivery, all orders must give either the *name of an individual* or an *official purchase order number.* Please submit your order as follows:

Subscriptions: specify series and year subscription is to begin.
Single Copies: specify sourcebook code (such as, PE8) and first two words of title.

Mail orders for United States and Possessions, Latin America, Canada, Japan, Australia, and New Zealand to:
Jossey-Bass Inc., Publishers
433 California Street
San Francisco, California 94104

Mail orders for all other parts of the world to:
Jossey-Bass Limited
28 Banner Street
London EC1Y 8QE

New Directions for Program Evaluation Series

Ronald J. Wooldridge, Ernest R. House, *Editors-in-Chief*

PE1 *Exploring Purposes and Dimensions,* Scarvia B. Anderson, Claire D. Coles
PE2 *Evaluating Federally Sponsored Programs,* Charlotte C. Rentz, R. Robert Rentz
PE3 *Monitoring Ongoing Programs,* Donald L. Grant
PE4 *Secondary Analysis,* Robert F. Boruch
PE5 *Utilization of Evaluative Information,* Larry A. Braskamp, Robert D. Brown

Contents

Editor's Notes

Internal evaluation is characterized by the use of internal staff or of contractors closely bound to the organization to conduct evaluation activities. The usual focus of internal evaluation is programs or problems of direct relevance to the organization's internal management. In contrast to external evaluations, those responsible for internal evaluations are often charged with remedying problems, not only with diagnosing them and developing recommendations.

Internal evaluation is growing rapidly. Disillusionment with the use of external evaluators, reductions in funding for large-scale evaluative research programs, and mounting concern for the management of human service organizations have contributed to this growth. As earlier generations of evaluators have assumed senior administration posts within organizations, they have stimulated the development of internal evaluation units. Increasingly, evaluators are finding a demand for their services as internal evaluators, some in the challenging role of evaluator-managers.

Despite this growth, there is still a serious gap in knowledge about the conduct of internal evaluation. The formal training of evaluators has left them unprepared for the precarious world of internal evaluation, particularly for the "action settings" of human service organizations. These organizations include service organizations spanning state and local government services, health services, education, social welfare services, and mental health services.

In the action settings provided by human service organizations, internal evaluation is nearly the converse of the description of a voyage around the globe offered by the sailor who called it "endless hours of sheer boredom, broken by periods of stark terror." That is, the internal evaluation of human service programs often involves endless hours of tossing and bobbing on the turbulent waters of these settings, interspersed with fleeting moments of reflective calm.

The skills involved in internal evaluation are both basic survival skills and the skills necessary for achieving superior results. Relatively little is known about evaluation that takes place within the organizational context. In Chapter One, Arnold Love reviews the unique characteristics of the organizational context and shows how internal evaluation fits within them. He delineates the major factors that define the organizational context, including the emergence of the evaluation resource management function.

That little is known about internal evaluation does not imply that internal evaluation does not have a long and honored history. Ancient tablets record that evaluation staff took samples of alcoholic brew, thereby providing perhaps the first quality assurance function for mankind! Internal evaluators have occupied a pivotal position in industry and government ever since then. The Roman *missi*, charged with monitoring the performance of Roman taxation, revenue, and military systems, were the forerunners of the auditors general and the inspectors general that nearly every civilized country in the world has today.

Modern internal evaluation is a child of the management review activities generated by the technology and information explosions of the twentieth century. The increasing complexity of business and management created a demand for improved mechanisms of planning and control. The work of Frederick Taylor and his successors on scientific management led to detailed studies of factory costs that laid the groundwork for modern cost and management accounting. Administrative management theory contributed to this development through its emphasis on control, departmental responsibility, and accountability.

The evolution of internal evaluation as a staff support service is directly linked with the creation of systems departments to provide expert knowledge for managerial problem solving. Systems staff successively extended their range of skills from mastery of administrative systems and procedures to industrial engineering techniques, market research methods, resource allocation procedures, planning and development activities, and decision–support systems. The internal evaluation function in business and industry is staffed by generalists who are expert both in technical domains and in all aspects of the corporate operation.

The picture for internal evaluation in human service organizations is less fortunate. The history of internal evaluation in human service organizations is both much shorter and decidedly less impressive. Moreover, instead of forming an elite team of troubleshooters, the evaluators in such settings are often seen as an isolated and aloof group of technicians who spend much of their time worrying about the legitimacy of their function. The image of the internal evaluator appears to fluctuate between that of a hatchet man for the executive director and that of an emissary from Babel who speaks in a strange tongue about incomprehensible topics. To rectify this situation, evaluation experts who are experienced human services managers or administrators have pooled their knowledge to provide answers in this volume about the how-to of internal evaluation in human service organizations.

In Chapter Two, Clifford and Sherman define the tasks of management and the complementary role of the evaluation professional. They outline the specific skills that the internal evaluator needs, and they identify the markers of success. Finally, they discuss the ethical issues that the internal evaluators must face.

Chapter Three also examines the interrelationship of evaluation and management activities. Focusing on the results of the interaction between these two professional disciplines, Neigher and Metlay investigate evaluation in practice and compare the competencies required for successful program evaluation and management. In an analysis of the respective reward and incentive systems for the program evaluation and management, the authors note that managers require reliable, quantitative techniques that can be used to provide dependable information for decision making. Program evaluators, they observe, have rarely been able to provide such information.

In Chapter Four, Newman and colleagues address the issue of data dependability. The authors argue that the dependability of clinical and fiscal data can be maintained at a level useful for internal evaluation if precautions are taken to control for organizational influences, data quality, staff concerns and staff and patient variables that bias the data. The first three areas are addressed in Chapter Four.

Chapter Five represents the culmination of many years of effort to increase data dependability and thereby remove a major obstacle to development of effective internal evaluation in human service organizations. In Chapter Five, Newman and colleagues present the results of several large-scale empirical studies aimed at identifying and controlling variables that affect the dependability of internal evaluation data. Empirical data that probe the data dependability issue are emphasized.

In Chapter Six, Landsberg describes a key tool for internal evaluation in human service organizations, the client utilization study, which makes information essential for decision making available to senior management, supervisors, and board members. Landsberg provides a checklist that measures an organization's ability to conduct such studies and illustrates various applications of this useful tool.

The last chapter in this volume is a reflective commentary by Wildavsky, the person whose name is most closely associated with self-evaluation. In Chapter Seven, Browne and Wildavsky address the implications of the fact that, in action settings, implementation and evaluation are often carried on by the same person, and they raise the question, Should evaluation become implementation?

Internal evaluation has enormous potential for the management of human service programs. However, if it is to fulfill this potential, significant change in the values and training of evaluators will be required. The maximum scientific rigor must be brought to the internal evaluation enterprise, as well as the maximum creativity. The contributors to this volume hope to stimulate attention to the development of effective internal evaluation.

Arnold J. Love
Editor

Arnold J. Love is a senior associate with Community Concern Associates Ltd. in Toronto, Canada. He specializes in the development of internal evaluation systems for human services organizations.

Organizational factors that influence the development of
effective internal evaluation are identified.

The Organizational Context and the Development of Internal Evaluation

Arnold J. Love

A flexible capacity for internal evaluation is fundamental to the management and ongoing improvement of a broad spectrum of organizations, including state and local governments, health and mental health services, social services, and education (Attkisson, Brown, and Hargreaves, 1978; Binner, 1975; Caplan, Morrison, and Stambough, 1975; Smith, 1972; Wildavsky, 1979). In times of economic turbulence, the internal evaluation process is essential for survival, because it provides the information crucial for program improvement, accountability, and planned change under adverse circumstances. The notion of the self-evaluating organization that uses program evaluation as the basis for program development and change, remains largely an ideal more than a decade after it was described by Aaron Wildavsky (1972). In that seminal work, Wildavsky described the types of barriers that obstruct the internal evaluation process. The major obstacles are not technological; instead, they are dilemmas posed by the organizational context of internal evaution. These dilemmas can be stated as questions: Who will evaluate, and who will administer? How will power be divided among administrators and evaluators? Can authority be allocated to evaluators and blame to administrators? How can managers be convinced to provide information that may help others but harm them?

A. J. Love (Ed.). *Developing Effective Internal Evaluation.* New Directions for
Program Evaluation, no. 20. San Francisco: Jossey-Bass, December 1983.

This chapter describes the fundamental importance of the organizational context for internal evaluation. In particular, it addresses the pivotal position that organizational context plays in determining the usefulness of evaluation information. Special attention is given to strategies for developing an organizational context that facilitates internal evaluation and constructive use of evaluative information in the turbulent action settings of human service organizations.

Evaluation in the Organizational Context

Let us start by examining the unique character and function of evaluation in the organizational context. This brief examination is necessary, because there is considerable debate and doubt about the legitimacy of program evaluation conducted by internal staff.

Distinguishing the major ways in which evaluation can be used provides a better understanding of the special uses of evaluation within the organizational context. Windle and Neigher (1978) have distinguished three models for the use of evaluation information: accountability, advocacy, and program improvement.

If information is used for accountability, this use implies that evaluation will focus on the effectiveness of programs in meeting the needs of their clients and on the quality of the services provided. Evidence of program impact can be compared with program costs to improve decisions concerning program support. The accountability function of evaluation can be used equally by policy makers and funders external to the organization and by managers and administrators internal to the program.

The external accountability role of evaluation emphasizes six areas: program outcomes, the causal relation between program activities and program outcomes, attainment of program goals, comparisons with similar programs, cost-effectiveness or cost benefits, and the consequences of policies, legislative compliance, or both. The evaluation methods used for external accountability are the methods traditionally identified with program evaluation: rigorous, objective, and independent measures of program effects by external evaluators using experimental designs and maintenance of tight control over program activities in order to comply with the evaluation design. Proper interpretation of evaluation results usually requires sophistication in the technical aspects of program evaluation.

The 1970s have been characterized as a decade of accountability for public sector and human services organizations. Within the space of ten years, program evaluation has been identified increasingly as a vehicle for external accountability. However, this is only one of several major functions that evaluation can fulfill. The use of evaluation for advocacy is grounded on the assumption that programs, agencies, and departments within organizations are competing for resources and that information is useful in such competition. When evaluation standards are high, information concerning program activities and outcomes can legitimately be used for advocacy. The advocacy use of program evaluation can be used inappropriately either to justify a weak program or to recommend the reform or elimination of a politically undesirable program.

The program improvement use of evaluation assumes that managers and professional staff want to improve the performance of their programs. In particular, given the changing and turbulent economic and political environment, they want to make better decisions about organizational and program options and about the use of scarce resources. If the bottom line of program evaluation remains the improvement of human services (Attkisson and Broskowski, 1978), achieving this aim depends on the use of evaluation information by decision makers.

The Shifting Meaning of Program Evaluation

Acknowledgement of the legitimacy of the use of evaluation for program improvement, advocacy, or internal accountability has profound implications for the method and conduct of evaluation. The primary concern for evaluation information shifts from government administrative agencies, legislatures, and the academic community to program managers and clients and indirectly to government agencies and legislatures. The responsiveness of evaluation to managers' information needs becomes a major consideration. Evaluation objectives and reports must become relevant to managers, service delivery staff, funding sources, and client groups. Evaluation must become sensitive to the rights of each group.

The shifting meaning of program evaluation is exemplified by the writings of Attkisson and Broskowski (1978). Building on the distinction made by Suchman (1967) and Weiss (1972) between program evaluation and evaluative research, these influential program evaluation specialists identify program evaluation as a tool for the management of organizations. According to these authors, "program evaluation must be viewed as an integral aspect of organizational design and organizational development. In other words, program evaluation is a function that takes place within an organizational context" (Attkisson and Broskowski, 1978, p. 22). In these authors' view, "evaluation is primarily a process within ongoing organizational management, decision making, and planning and is only secondarily a research enterprise for the purpose of new scientific discoveries" (p. 22). This conceptualization of program evaluation marks a clear departure from the commonly held view that program evaluation is the application of social science methodologies, particularly of randomized controlled experiments, to the measurement of program outcomes (Rossi and Wright, 1977).

Human services organizations are just beginning to respond to the requirement for sounder management (Drucker, 1980; Stretch, 1978). Organizational survival requires that service organizations establish and document the relevance, effectiveness, and efficiency of their services. They must manage for results by eliminating programs that duplicate or overlap those of other community services. A comparative study by van de Vall (1975) supports the perception that internal evaluation is widely used in business and industrial settings but not in human service organizations. In a study of 120 organizations, van de Vall found that 58 percent of the industrial organizations used internal evaluation for their evaluation studies. External evaluation was used for only 15 percent of their projects. In human service organizations, the opposite was the case. Internal evaluation was used for only 8 percent of the projects, and external evaluation was employed for 70 percent of the projects.

The Two Major Functions of Internal Evaluation

Internal evaluation systems are intended to influence decision making and organization performance in two major ways: first, by providing relevant and timely information on which managers can base decisions; second, through the evaluation process itself, which has certain effects on decisions and performance as well as on the decision-making process. Flamholtz (1979) has termed the first of these functions the informational function and the second the process function. In the organizational context, the evaluation process has the dual purpose of providing information and of influencing behavior, including decision-making behavior. This distinction implies that a different set of criteria must be used to evaluate the effectiveness of internal evaluation systems: the criteria of behavioral validity and behavioral reliability, in addition to the criteria of statistical validity and reliability.

The primary characteristic that distinguishes evaluation processes within the organizational context is the degree to which they are intended to influence human behavior as well as to measure and represent the attributes of objects. The very existence of a measurement system can have effects on human behavior in an organization regardless of the numbers that are the outputs of that system (Flamholtz, 1979). Although the notion of the behavioral consequences of evaluation is not new, the evaluation literature has treated behavioral responses as unintentional aberrations or pathologies (Guba, 1969; Weiss, 1973). Analogies have linked the organizational response to evaluation with neurotic pathology characterized by anxiety, avoidance, immobilization, fear of failure, and similar dysfunctional behaviors.

The behavioral consequences of evaluation should be viewed as an integral component of the evaluation process. Too much emphasis has been placed on the technical adequacy of evaluative measures, while the fact that the evaluation process functions in ways that are both more diverse and more complex than the traditional criteria of statistical validity and reliability can capture has received too little attention.

The Influence of Organizational Context
on Internal Evaluation

The organizational context, including its structure and role relationships, strongly influences both the type of evaluation that is conducted and the use of its results (Broskowski and others, 1979; Chapman, 1976; Conner, 1979; Gurel, 1975; and Sjoberg, 1975). A deeper understanding of the behavioral aspects of internal evaluation systems is essential for anyone who is concerned with the intelligent management of organizational change and improvement (Hopwood, 1974). Once the psychological and behavioral aspects of internal evaluation systems are recognized, organizational strategies that enhance the effectiveness of the self-evaluation process can be examined and implemented. Despite the importance of the organizational context for internal evaluation, no discrete body of knowledge explicating this relationship exists (Cohen, 1977). If the ideal of the self-evaluating organization is to be achieved, steps need to be taken to improve this situation.

The circuitously complex and continually shifting nature of human service organizations has presented a persistent challenge to internal and external evaluators alike. Each organizational setting appears to be unique. Each organization possesses a different mission, a different constituency, a different range of services, a different type of organizational structure, and different purposes for evaluation. A promising approach to the study of organizational context lies in the identification of a comprehensive set of variables affecting the success and failure of information feedback and evaluation in organizations (Broskowski and Driscoll, 1978; Ein-Dor and Segev, 1978; Lucas, 1973; Rothman, 1980). The interdependency in modern organizations between the internal evaluation function and its information systems (Attkisson, 1980; Attkisson, Hargreaves, and others, 1978) makes the investigations of Ein-Dor and Segev (1978) particularly relevant.

Variables That Affect the Success of Internal Evaluation

Ein-Dor and Segev (1978) present a useful conceptual scheme and categorization of organizational context variables that affect the use of information by managers. The variables are characterized as uncontrollable, partially controllable, and controlled. The scheme is useful both for assessing the feasibility of developing an internal evaluation function and for analyzing problems or feasibility of change in functioning systems.

According to these authors, uncontrollable variables are those not under direct control of the organization or those not amenable to change within a reasonable period of time. The uncontrollable variables are organizational size, organizational structure, organizational time frame, and the extraorganizational situation. The partially controllable variables are those that can be changed in the desired direction during a reasonable period of time. The partially controllable variables are organizational resources, organizational maturity, and the organization's psychological climate. Ein-Dor and Segev suggest that in many cases it may be possible to modify the aspects of the organizational context that are partially controllable either before or during the implementation of an information feedback system. The third kind of variable is completely under the control of the organization, and it can be determined by the organization with precision at any time. There are two of these fully controllable variables: the rank and location of the executive responsible for evaluation and the steering committee. Suggested operational measures for these organizational context variables are presented in Table 1. The original suggestions offered by Ein-Dor and Segev have been modified somewhat to make the measures more relevant to human service organizations. Even so, the exact measures used in any particular case would require matching to the specific organizational situation.

In addition to the conceptual scheme, Ein-Dor and Segev (1978) present a series of propositions concerning each variable and the interactions among them. Although the propositions are advanced as a basis for research designed to formalize the currently intuitive approach that prevails, they provide a useful framework for categorizing and defining strategies that can be used to develop or modify organizational context so as to render it consonant with successful implementation. The strategies have been compiled from numerous sources, including Attkisson, Brown, and Hargreaves (1978), Broskowski and Driscoll (1978), Cameron (1978), Courey (1978), Ein-Dor and Segev (1978), Kimmel (1981), Likert (1967), Love and Shaw (1981), and Rothman (1980). The propositions have been organized into four broad areas: organizational support, organizational structure, psychological climate, and evaluation resource management. In the sections that follow, the interrelationships between variables and the strategies related to them will be described briefly.

Organizational Support. Four major factors are associated with organizational support for internal evaluation: policy, funding, skilled staff, and time frame.

An organization that has an expressed policy to undertake and support internal evaluation provides visible proof of the commitment of the board of directors and senior management to internal evaluation. A clearly stated policy regarding evaluation legitimizes the evaluation function, promotes positive expectations and attitudes about evaluation throughout the organization, provides support for internal evaluation at the highest levels of the organization, and facilitates resource allocation for evaluation.

The presence of adequate and stable funding is essential for developing the internal evaluation function. In times of economic constraint, internal evaluation must compete with other functions for scarce resources. Because internal evaluation is often perceived as optional, rather than as an integral

Table 1. Operational Measures for Organizational Context Variables

Organizational Context Variable	Operational Measures
Organizational size	Number of clients or patients annually; size of staff; assets; client capture area
Organizational structure	Number of divisions; number of departments; number of hierarchical levels; lines of authority and responsibility
Organizational time frame	Planning horizon; average length of strategic decision process; rate of legislative and funding changes; rate of technological and professional skills change in the service sector
Extraorganizational situation	Availability of skilled evaluation staff; availability of information system hardware and software; availability of decision-relevant data
Organizational resources	Size of budget; size and type of skilled staff; financial position
Organizational maturity	Degree of system formalization; level of qualification; availability of decision-relevant data
Psychological climate	Attitudes to evaluation; perceptions of evaluation; expectations for evaluation systems
Rank of executive responsible for evaluation	Number of levels below executive director
Location of executive responsible for evaluation	Identification with specific functional area, planning or administrative services
Evaluation/information systems steering committee	Existence; organizational level

part of management, it often loses in such competition. Lack of reliable fiscal resources has been a frequent contributor to failure of internal evaluation in human service organizations. The level of resources required for internal evaluation, usually 3 to 5 percent of the annual budget, has presented a serious problem for smaller organizations. The usual result is in understaffing or haphazard evaluation activities. This fact favors the implementation of internal evaluation in larger organizations.

Adequate staffing by skilled evaluation staff is another prerequisite for the success of internal evaluation. Sadly, adequate staffing is the exception rather than the rule — for several reasons. Usually persons skilled in internal evaluation or who have formal training in both evaluation and management are difficult to recruit, and persons with little knowledge of evaluation have often been pressed into service. These persons range from part-time clerks, bookkeepers, or records personnel to academic researchers and management consultants. Nevertheless, the rapid shift from emphasis on large-scale external evaluation projects by government and the emphasis on management of human services organizations have created a sharp demand for internal evaluators. The demand still outstrips the supply, forcing human service organizations to recruit and train their own internal evaluation staff.

The hallmark of internal evaluation is the systematic use of evaluative information for decision making. Perhaps the most significant step in the evolution of internal evaluation capability comes when an organization turns from gathering information for a specific application or a single end-user's needs to a structured system providing ongoing information to all decision makers. A sufficiently long time frame is required for this development, including a firm commitment of staff and resources to the process.

Organizational Structure. The structure of an organization has repeatedly been identified as fundamental to the successful development and implementation of internal evaluation and management information systems. Structural factors are crucial to internal evaluation because they influence both the flow and processing of information and the behavior of the organization. Because these are two major aspects of internal evaluation, structural variables have a major impact on the use of internal evaluation information. The factors selected for inclusion under organizational structure — organizational maturity, evaluation capability, role definitions, and location of internal evaluation unit — reflect this dual nature of internal evaluation.

Organizational maturity refers to the presence of formalized systems that provide information appropriate to their decision and control processes. Maturity, therefore, is measured by the presence of a rational structure and by the use of information to guide the direction of the organization and its operation. The type of organizational structure, which ranges from a functional organization based on clearly delineated lines of authority and functional responsibility to a systems structure that uses team approaches and decentralization, varies with the complexity and mission of the organization. An organization can be mature (or immature) irrespective of its type of organization.

The development of evaluation capability is strongly related to organizational maturity. The evolution of evaluation capability, both in terms of evaluative activity and of information capability, has been described by Attkisson, Brown, and Hargreaves (1978), and it will not be reviewed here. It is sufficient to say that effective internal evaluation depends on timely access to reliable and valid data that are relevant for management decision making. The higher levels of evaluative capability are more adequate in providing this information, and these higher levels are difficult to achieve unless the organization has formal mechanisms for providing quantitative data on a regular basis. Also, data from diverse sources, such as client records, budgets, and personnel records, must be accessible for use. Likewise, the higher levels of evaluation capability are not likely to evolve unless the social and political structure of the organization supports a data-based approach to management decision making. Smaller organizations may require considerable time and effort to formalize and systematize the organization before higher levels of evaluation capability can be attained. This does not mean that smaller or less formalized organizations should not begin internal evaluation. In fact, successful small-scale evaluations tend to enhance the level of organizational maturity. The interrelationship between organizational maturity and evaluation capability should be recognized.

The next factor in organizational structure—the role definitions of evaluators and managers—has received much attention in the evaluation literature, and these definitions are detailed in Chapters Two and Three of this volume. One common thread in the literature is that the evaluator can have many different roles in the organization and that the effectiveness of these roles is closely linked to the structural connection between evaluators and other members of the organization (Rothman, 1980). Structural linkages include the relationship between evaluators and senior management, middle management, and frontline staff; the existence of a formal internal evaluation function; the size of the evaluation unit; the membership and skills of the evaluators; the lines of authority and responsibility of evaluators; and the lines of communication between evaluators, program managers, and other members of the organization.

The final structural factor is the location of the internal evaluation unit within the organization. Location includes three elements: the function of the evaluation unit within the organization, the rank and reporting relationships of the executive responsible for the evaluation unit, and the relationship between the evaluation unit and other units of the organization.

Internal evaluation units in industrial settings have a staff support function. In most cases, they are a troubleshooting adjunct to senior management. In these settings, internal evaluators are responsible both for diagnosing problems and for correcting them. This heavy responsibility requires persons who are experts in their fields and who have an intimate knowledge of organizational systems and operations. In fact, internal evaluation units more often resemble think tanks or consulting firms peopled by expert generalists who have risen from the ranks. There is usually a direct link to planning functions. In contrast, internal evaluation units in human services organizations are often dissociated from any particular function, and they often operate as separate entities. Usually, these units do not maintain regular contact with program managers and frontline staff, which serves to impart a sense of isolation and perceived exclusiveness. The evaluators tend to adopt the roles of clerks, statisticians, or researchers rather than of management consultants.

The remedy seems to involve three strategies. First, the internal evaluation executive should report to the highest levels in the organization, preferably to the executive director or the assistant executive director. There should be a visible identification with the highest echelons in the organization, and internal evaluation should receive vocal support from senior management. Second, the internal evaluation function should be linked to planning and management activities. Evaluators should sit on central planning committees and on committees involved in program planning. Third, the role of evaluator should be clearly defined as that of adviser or consultant to program managers. The linkage between managers and evaluators should be established both formally and through informal communication channels. The authority and responsibility for evaluation should clearly rest with senior managers and program managers. Evaluators should serve in an advisory role with respect to both the technical and the behavioral aspects of evaluation. Evaluators should participate actively with program managers in the process of using evaluation information.

Psychological Climate. Each of the major organizational theories proposes a specific conceptualization of optimal organizational functioning, together with a set of principles and processes necessary to achieve it. The different theoretical positions reflect various assumptions about human nature, the economic basis for human organization, the integration of individual goals and the goals of the organization, and the control of human behavior within the organization.

Gurel (1975) and Sjoberg (1975) have stated the "First Law of Evaluation": Evaluations of ongoing programs produce resistance to recommended changes regardless of their size, methods, or sponsorship. Not all types of evaluation generate equal amounts and kinds of resistance. However, this unequal resistance is less likely to emanate from differences in the type of evaluation than it is from the structural restraints and requirements, as well as the interpersonal relationships, that characterize the organizational environment (Bonoma, 1977).

The evaluation literature is replete with reports that link the psychological climate of organizations with the success of their internal evaluation efforts (Attkisson and Browskowki, 1978; Bonoma, 1977; Chapman, 1976; Gurel, 1975; Rothman, 1980). Specific types of organizations appear to possess psychological climates that nurture the evolution of self-evaluation. In a study of organizational climates, Likert (1967) described four types of organizations: exploitive-authoritative, benevolent-authoritative, consultative, and participative. The major characteristics of these four types of organizations are compared in Table 2.

The correspondence between the participative type of organization and the ideal psychological climate for internal evaluation is clear from Likert's (1967, pp. 135–136) description of the purpose of measurement and information systems in the participative system:

1. The primary purpose is to provide managers and nonsupervisory employees with information to help guide their own decisions and behavior.

2. Information is used to help guide decisions and actions, and [it is] not used punitively by superiors.
3. All members of the organization want the data and clearly recognize the necessity for all the measurements to be accurate.
4. There are strong motivational forces among all members to ensure the data are accurate and valid because the measurements are used for self-guidance.
5. The high levels of confidence and trust enable accurate measurements once obtained to flow to all relevant parts of the enterprise to provide correct information to all persons who have need for it.

Efforts to turn an organization toward the participative context need to deal with the organizational factors that bind the organization to its management system. These factors include philosophies, values, standards, methods of decision making, lines of authority and responsibility, and use of resources. The policies, values, and philosophies of the participative organization encourage supportive behavior throughout the organization. Information is shared, and it is used for problem solving throughout the organization. Workers are trusted. Clear communication is actively encouraged. High standards of quality are maintained through regular evaluation of results. Loyalty to the work group and the organization is fostered. The contributions of workers and management are valued.

Evaluation Resource Management. Information is the common denominator of effective communication and control within organizations regardless of size, scope of operations, or private or public sector affiliation. However, the usual methods for developing internal evaluation systems have emphasized specific user applications and data processing. The idea of managing evaluation as a corporate resource, as an activity distinct and separate from the management of individual evaluation projects, is emerging rapidly as the importance of evaluation to overall organizational management is recognized. The major question for human service organizations is how evaluation information can be managed as a central resource while also remaining responsive to the specific applications of individual program managers. Because the concept of evaluation resource management is so recent, the approaches suggested here should be viewed as interim methodologies until more firmly established strategies become available.

Internal evaluation has been used primarily for operational control in human service organizations. Recently, there has been increased demand for evaluation information for management control and for strategic planning purposes. Operational control assures that specific tasks — for example, the admission of patients to an emergency ward, the use of a clinical intervention — are carried out efficiently and effectively. Management control assures that resources are obtained and used effectively and efficiently to accomplish the organization's objectives. In contrast, strategic planning is used to decide and

Table 2. Four Types of Organizations

Organizational Factors	Authoritative Exploitive	Authoritative Benevolent	Consultative	Participative
How much confidence is shown in subordinates?	None	Condescending	Substantial	Complete
How free do subordinates feel to talk to superiors about job?	Not at all	Not very	Rather free	Fully free
Are subordinates' ideas sought and used if worthy?	Seldom	Sometimes	Usually	Always
Is predominant use made of (1) fear (2) threats (3) punishment (4) rewards (5) involvement?	1, 2, 3, occasionally 4	4, some 3	4, some 3, some 5	5, 4 based on group-set goals
Where is responsibility felt for achieving the organization's objectives?	Mostly at the top	At the top and middle	Fairly general	At all levels
How much communication is aimed at achieving the organization's objectives?	Very little	Little	Quite a bit	A great deal
What is the direction of information flow?	Down	Mostly down	Down and up	Down, us, and sideways
How is downward communication accepted?	With suspicion	Possibly with suspicion	With caution	With an open mind
How accurate is upward communication?	Often wrong	Censored for the boss	Of limited accuracy	Accurate
How well do superiors know problems faced by subordinates?	Very little	Somewhat	Quite well	Very well
At what level are decisions made?	Mostly at the top	Policy at the top, with some delegation	Broad policy at the top, with more delegation	Throughout, but well integrated

What is the origin of technical and professional knowledge used in decision making?	Top management	Upper and middle management	To a certain extent throughout	To a great extent throughout
Are subordinates involved in decisions related to their work?	Not at all	Occasionally consulted	Generally consulted	Fully consulted
What does the decision-making process contribute to motivation?	Nothing, often weakens it	Relatively little	Some contribution	Susbstantial contribution
How are organizational goals established?	Orders are issued	Orders are issued; some comment is invited	After discussion, by orders	By group (except in crisis)
How much covert resistance to goals is present?	Strong resistance	Moderate resistance	Some resistance at times	Little or no resistance
How concentrated are review and control functions?	Highly at top	Relatively high at top	Moderate delegation to lower levels	Quite widely shared
Is there an informal organization resisting the formal one?	Yes	Usually	Sometimes	No; same goals as formal
What are cost, productivity, and other control data used for?	Policing, punishing	Reward, punishment	Reward, some self-guidance	Self-guidance, problem solving

Source: Likert, 1967.

change the major goals of the organization, to determine the programs and resources needed to achieve specific objectives, and to define policies that govern the use of these resources.

The private sector has been using internal evaluation and information systems for management support and strategic planning in coping both with the recession and with market instability. The human services sector is only now beginning to develop capability in these two areas. It is not surprising, therefore, that internal evaluation in human service organizations has focused on specific operational or management problems or that it has responded to specific applications or to the information needs of a single manager or management group. Concentration on specific evaluation applications has produced disjointed evaluation systems, with attendant problems of redundancy, lack of data independence, and an inability to meet ad hoc information requests. The absence of integration of the evaluation function jeopardizes the management control and strategic information needs of the total organization.

Internal evaluation in the human services usually has been organized around project teams or individual evaluators working directly with a program manager. Individual project teams or evaluators are seldom aware of the relationship of functions and information across boundaries within the organization or between specific projects, and if they are aware of it, they have neither the time nor the authority to address such issues. The needs of alternate users of evaluation information or the relationship to information in other areas of the organization are rarely addressed.

Two parallel but distinct sets of activities are required to develop effective internal evaluation systems: activities related to the processing of required information and to the relating of these information needs to the information needs of other functions in the organization, and activities related to understanding the information needs of specific users and to the design and implementation of evaluation studies that can provide that information. The heart of the problem is that existing internal evaluation strategies fail to recognize that there is a crucial difference in the management of these two sets of activities. The activities related to user applications can continue to be carried out by individual evaluators or project teams in a personalized and decentralized manner, but the absence of centralized control thwarts integration. It is necessary, therefore, to evolve a function that has specific responsibility for evaluation resource management across all application projects.

Evaluation resource management is designed to improve the availability, accessibility, and use of evaluative information within the organization. An effective evaluation resource management function has the five objectives: It assists in the efficient development of specific applications that support operations, management, and control. It provides support for strategic decision making by making relevant information available for managers. It provides flexible access to information. It centralizes the evaluation data base while insulating the data from the processing required to meet specific user needs.

And, it develops and maintains quality standards for evaluation studies and evaluation information. These objectives are achieved through the execution of two principal functions: data administration, which manages and controls the definitions, relationships, and documentation of the organization's evaluation information, and data base management, which provides the facilities to store, modify, and access the organization's evaluative data.

Data Administration. Data administration assists in the management of evaluation resources by taking responsibility for developing an organizational data model, by collecting information about data in the organization and storing this information in an easily accessible form (data dictionary), by developing an applications data model, and by developing standards and procedures for the storage, access, and use of evaluation information. Because the concepts of data dictionaries and standards are well understood, I will address the two crucial elements of data administration, namely, the organizational data model and the application data model.

The organizational data model provides a top-down view of the organization's data, especially evaluation data. The model is necessary to obtain a comprehensive picture of key organizational functions, processes, activities, and data relationships within the organization. The model permits the definition of subject data bases, which are relatively small sets of related data bases. The model also permits the verification of the applications data model.

The application data model is created bottom-up through close contact between evaluators and the organization's frontline staff. This model is very detailed, and the information represents the data entities (such as clients, services, records) and the data elements (such as client number, service contact number, record number) as well as the relationships between them. Further, the applications data model is verified at the lowest level of user in the organization (usually frontline staff).

The requirements of an evaluation project team or individual user rarely make use of the entrie applications data model. The subset of data elements and relationships of interest for a specific application is called the user view. For example, frontline clinical staff may want information about the effectiveness of a certain treatment process for a certain type of client. In contrast, program managers may want to know the relative cost-effectiveness of several treatment processes for different types of clients. The user view is somewhat different in each case.

The development of the application data model helps to ensure ownership of the evaluation process by the persons who must collect data (usually frontline staff) thereby increasing involvement and reducing resistance to evaluation. Also, it attempts to project the organization's future information needs and thereby meet the needs of applications that have yet to be identified.

Data Base Management. Data base management activities and the roles of persons who perform them are well known. Primarily, their task is to translate the applications data base design developed by the data administration function into a form usuable by the data base software. Procedures to protect the confidentiality of records, to provide back-up for the data base, and to provide user access routines are also part of this function.

Future Directions

During the next few years, we can expect the growth and change in the internal evaluation function to continue. The interrelationship of evaluation and the organizational context is likely to become more widely recognized. A more systematic approach to the development of internal evaluation capability, coupled with empirical studies, will increase the usefulness and success of internal evaluation implementations. The character of internal evaluation is likely to move from single user applications and operations studies to a wider focus on management control and strategic planning. The organization of the evaluation resource function will become more formalized as internal evaluation matures.

Obstacles to the growth of internal evaluation will continue. Some of these are addressed in the other chapters of this sourcebook. The roles of evaluators and managers require further clarification, the skills of the internal evaluator require sharper definition, and the purposes of internal evaluation demand legitimatization, while the questions of data dependability and the effects of bias remain persistent issues. Despite these obstacles, I believe that internal evaluation will become an important part of human services management and that it has the potential to foster the development of more efficient, more effective, and more responsive organizations.

References

Attkisson, C. C. "The Manager as Evaluator." *New Directions for Mental Health Services,* 1980, *8,* 77–90.
Attkisson, C. C., and Broskowski, A. "Evaluation and the Emerging Human Service Concept." In C. C. Attkisson, W. A. Hargreaves, M. J. Horowitz, and J. E. Sorensen (Eds.), *Evaluation of Human Service Programs.* New York: Academic Press, 1978.
Attkisson, C. C., Brown, T. R., and Hargreaves, W. A. "Roles and Functions of Evaluation in Human Service Programs." In C. C. Attkisson, W. A. Hargreaves, M. J. Horowitz, and J. E. Sorensen (Eds.), *Evaluation of Human Service Programs.* New York: Academic Press, 1978.
Attkisson, C. C., Hargreaves, W. A., Horowitz, M. J., and Sorensen, J. E. (Eds.). *Evaluation of Human Service Programs.* New York: Academic Press, 1978.
Binner, P. "Program Evaluation." In S. Feldman (Ed.), *The Administration of Mental Health Services.* Springfield, Ill.: Thomas, 1975.
Bonoma, T. V. "Overcoming Resistance to Changes Recommended for Operating Programs." *Professional Psychology,* 1977, *8* (4), 451–463.
Broskowski, A., and Driscoll, L. "The Organizational Context of Program Evaluation." In C. C. Attkisson, W. A. Hargreaves, M. J. Horowitz, and J. E. Sorensen (Eds.), *Evaluation of Human Service Programs.* New York: Academic Press, 1978.
Broskowski, A., White, S. L., and Spector, P. E. "A Management Perspective on Program Evaluation." In H. C. Schulberg and J. M. Jerrell (Eds.), *The Evaluator and Management.* Beverly Hills, Calif.: Sage, 1979.

Cameron, K. "Measuring Organizational Effectiveness in Institutions of Higher Learning." *Administrative Science Quarterly,* 1978, *23,* 604-632.

Caplan, N. "A Minimal Set of Conditions Necessary for the Utilization of Social Knowledge in Policy Formulation at the National Level." In C. H. Weiss (Ed.), *Using Social Research in Public Policy Making.* Lexington, Mass.: Heath, 1977.

Caplan, N., Morrison, A., and Stambough, R. *The Use of Social Science Knowledge in Policy Decisions at the National Level: A Report to Respondents.* Ann Arbor: Institute for Social Research, University of Michigan, 1975.

Chapman, R. L. *The Design of Management Information Systems for Mental Health Organizations: A Primer.* DHEW Publication No. ADM76-333. Washington, D.C.: U.S. Government Printing Office, 1976.

Cohen, L. H. "Factors Affecting the Utilization of Mental Health Evaluation Research Findings." *Professional Psychology,* 1977, *8,* 526-534.

Conner, R. F. "The Evaluator-Manager Relationship: An Examination of the Sources of Conflict and a Model for a Successful Union." In H. C. Schulberg and J. M. Jerrell (Eds.), *The Evaluator and Management.* Beverly Hills, Calif.: Sage, 1979.

Courey, C. J. "Approaches to Assessment and Evaluation in a University Environment." *Managerial Planning,* 1978, *27* (3), 17-23.

Drucker, P. F. *Managing in Turbulent Times.* New York: Harper & Row, 1980.

Ein-Dor, P. and Segev, E. "Organizational Context and the Success of Management Information Systems." *Management Service,* 1978, *24,* 1064-1077.

Flamholtz, E. G. "Toward a Psychotechnical Systems Paradigm of Organizational Measurement." *Decision Sciences,* 1979, *10* (1), 71-84.

Guba, E. G. "The Failure of Educational Evaluation." *Educational Technology,* 1969, *9,* 29-38.

Gurel, L. "The Human Side of Evaluating Human Services Programs: Problems and Prospects." In M. Guttentag and E. L. Struening (Eds.), *Handbook of Evaluation Research.* Vol. 2. Beverly Hills, Calif.: Sage, 1975.

Hopwood, A. *Accounting and Human Behavior.* Englewood Cliffs, N.J.: Prentice-Hall, 1974.

Kimmel, W. A. *Putting Program Evaluation in Perspective for State and Local Government.* Washington, D.C.: Project Share Management Series, 1981.

Likert, R. *The Human Organization.* Toronto: McGraw-Hill, 1967.

Love, A., and Shaw, R. C. *Impact Evaluation.* Toronto: Dellcrest Children's Center, 1981.

Lucas, H. C., Jr. "A Descriptive Model of Information Systems in the Context of the Organization." *Data Base,* 1973, *5* (5-3-4), 27-39.

Rossi, P. H., and Wright, S. R. "Evaluation Research: An Assessment of Theory, Practice, and Politics." *Evaluation Quarterly,* 1977, *1,* 5-51.

Rothman, J. *Using Research in Organizations: A Guide to Successful Application.* Beverly Hills, Calif.: Sage, 1980.

Sjoberg, G. "Politics, Ethics, and Evaluation Research." In M. Guttentag and E. L. Struening (Eds.), *Handbook of Evaluation Research.* Vol. 2. Beverly Hills, Calif.: Sage, 1975.

Smith, G. "Research Policy: 'Pure' and 'Applied.'" *British Hospital Journal and Social Science Review,* April 1972, pp. 723-724.

Stretch, J. J. "Increasing Accountability for Human Service Administrators." *Social Casework,* June 1978, pp. 324-330.

Suchman, E. A. *Evaluative Research: Principles and Practice in Public Service and Social Action Programs.* New York: Russell Sage Foundation, 1967.

van de Vall, M. "Utilization and Methodology of Applied Social Research: Four Complementary Models." *Journal of Applied Behavioral Science,* 1975, *11* (1), 14-38.

Weiss, C. H. (Ed.). *Evaluating Action Programs.* Boston: Allyn & Bacon, 1972.

Weiss, C. H. "Evaluation in the Political Context." Paper presented at the annual meeting of the American Psychological Association, Montreal, August 1973.

22

Wildavsky, A. "The Self-Evaluating Organization." *Public Administrative Review,* 1972, *32,* 509–520.

Wildavsky, A. "The Self-Evaluating Organization." In H. C. Schulberg and F. Baker (Eds.), *Program Evaluation in the Health Fields.* Vol. 2. New York: Human Sciences Press, 1979.

Windle, C., and Neigher, W. "Ethical Problems in Program Evaluation: Advice for Trapped Evaluators." *Evaluation and Program Planning,* 1978, *1* (2), 97–107.

Arnold J. Love is a senior associate with Community Concern Associates Ltd. in Toronto, Canada. He specializes in the development of internal evaluation systems for human services organizations.

The roles and skills of internal evaluators, together with the markers of success, are discussed.

Internal Evaluation: Integrating Program Evaluation and Management

David L. Clifford
Paul Sherman

The internal evaluator is an employee of an organization who holds explicit, primary responsibility for the organization's self-evaluation. The key terms here are *internal, employee,* and *explicit responsibility.* These three characteristics distinguish the internal evaluator from the consultant, the academic-based evaluation researcher, and the funding agency evaluator or monitor. The pressures, and objectives that these characteristics create for the internal evaluator are very different from those of colleagues who function outside the organization. This chapter, which is based on our firsthand experience of years as internal evaluators, reviews some of these differences.

The internal evaluator's basic function is to support planning and management in such a manner as to maximize the amount of intelligence and analytical discipline involved in the decision-making process. Internal evaluation is a tool of management science as much as or more than it is either a product or a tool of social science. The internal evaluator has a long-term commitment to change through enhancement of the quality of decision making in the organization. The ideal approach is nonadversarial. In style and technique, the internal evaluator is formative, not summative.

A. J. Love (Ed.). *Developing Effective Internal Evaluation.* New Directions for Program Evaluation, no. 20. San Francisco: Jossey-Bass, December 1983.

In this chapter, we describe the role of the internal evaluator in the management and operations of human service organizations. We seek to explicate the skills required, the approaches involved in marrying evaluation and management, the markers of success for the internal evaluator, and some of the ethical dilemmas that he or she faces.

Tasks of Management in Human Service Organizations

Management in human service organizations has many tasks: obtaining the resources needed to fund the operation, deciding what the organization is to accomplish, staffing the operation and assigning responsibility and authority, allocating the resources available in order to accomplish the organization's objectives, devising systems for determining whether the organization is accomplishing what it intended to accomplish and, if it is not, making corrective interventions to restore the organization to its intended course. Management also has the responsibility to analyze the environment to determine what changes in its mission and operations will be needed to cope with changes occurring or likely to occur in that environment. Management theorists (Anthony and Herzlinger, 1980) say that performance (real-time feedback on operations) as well as operations and program evaluation are integral parts of the normative management process. This is true for all levels of management performance: strategic (policy making), management (effective and efficient use of resources), and operations (production or service delivery) (Anthony and others, 1965). Finally, management has the ethical responsibility to determine whether the organization is having unintended adverse effects on its environment and, if it is, to take corrective action.

While evaluation is a crucial aspect of all these management functions, our focus will be on the management functions of planning and controlling. As we use the term, *planning* relates to seeing both opportunities and threats in the future and taking steps in the present to exploit the opportunities and combat the threats (Starling, 1977). In our conceptualization, the internal evaluator has as much responsibility for supporting the planning function as for supporting the controlling function.

In order to perform its planning and control functions, management requires support from a variety of specialists, including behavioral scientists, operations researchers, statisticians, computer programmers, accountants, and planners. In most human service delivery agencies, evaluators are called on to play many of these roles concurrently.

Partitioning Management and Evaluation Roles and Responsibilities. What are the roles of manager and evaluation specialist in an organization where the evaluator has a decision-support function? In this section, we will describe the primary responsibilities of each party. We do not imply that these parties perform their functions in isolation.

It is the manager's responsibility to define the issues or problems of interest. The issues are management decision dilemmas. It is the manager's responsibility to make the authoritative decisions regarding the arena of potential options for response to the problem. The motivation for seeking answers is to reduce the uncertainty or risk involved in making a decision.

For example, let us assume that the management dilemma involves what should be done about decreasing the length of time that clients wait to receive services after their initial interview. The manager has given the problem some thought, and he or she has decided that it is not possible, given current budget and economic constraints, to hire additional staff, or to reallocate current staff. Six interventions can be considered viable: encouraging more referrals to other agencies, increasing referrals to self-help networks within the community, pressing agency staff to increase their use of the group treatment modality, decreasing the current length of stay for clients in treatment at the agency, some combination of the preceding, or doing nothing at all. The data-based evaluation-oriented decision-support approach to this dilemma involves specifying a series of questions, the answers to which will enable the manager to reduce his uncertainty about the costs, benefits, and consequences of the individual options. Some of these options may have undesirable effects, and it would probably be too costly to implement them all. Thus, the manager must evaluate each one. To evaluate option three, the manager can ask such questions as these: What proportion of the caseload is currently being treated in the group modality? What proportion of the caseload not being seen in a group modality could be served in such a modality without adverse effects? What is staff's attitude about increasing the proportion of cases treated in group modality if data on client characteristics suggest that more group treatment is justifiable? Is there a reason to believe that the net gain from treating clients in a group modality would relieve the backlog over the long term? It is obvious that switching to group modality would relieve the backlog over the short term, but if lengths of stay are allowed to increase for group clients, as experience predicts that they will, then there may be no long-term net gain.

The evaluation specialist and the manager have the joint responsibility of determining what data are needed to answer the question that the manager asks. While the primary responsibility rests with the evaluation specialist, there is great need for two-way communication between the manager and the evaluator in the early stages of the task. One of the issues that both must consider is what to do about questions that cannot be answered with existing data. Are the available data an adequate approximation of what is needed? If they are not, new data will have to be collected. This raises the issue of timeliness, cost, and value of additional information. The evaluator then has the responsibility for determining the analyses that need to be performed and the best format for presentation of the results. It is the manager's responsibility to consider the conclusions suggested by the data and then to decide on the course of action to pursue. The manager's decision may or may not be supported by the data. Political considerations can outweigh the data considerations.

Management Style. Many factors are involved in promoting a decision-support role for internal evaluators who work in human service organizations. Many of these factors will be addressed later in this chapter. At this point, it is appropriate to review the literature on managers in order to appreciate philosophical and methodological changes that must be made in management's typical style of operation in order to make the decision-support approach standard operating procedure.

One of the most comprehensive studies of management behavior (Mintzberg, 1973) depicts managers as people who do not have well-articulated models of how their organizational systems work. Managers tend to process information sequentially, and there is a strong recency effect. That is, the most recent pieces of information to be processed have a disproportionate impact on their decisions. Managers tend to rely heavily on oral input and to assign "decision worth" to information as a function of its source, not of the methodology used to collect it. Although this picture of management behavior is gloomy, since it suggests that attempts to institute a data-based approach to decision making will meet with resistance, much can be done to change it, as we discuss later in this chapter.

Identity and Skills for Internal Evaluators in Human Service Organizations

Identity. To function as a decision-support person in a human service organization, the evaluator must be a member of the management team. While this identity carries with it the threat that the evaluator will lose the objectivity traditionally associated with the evaluator's role, it is crucial to functioning as an internal evaluator. The evaluator's self-perception is also crucial. The evaluator can perceive himself or herself as a social researcher with some management skills or as a manager with a social research, planning, and other evaluation-related skills and experiences. If the aim is to institute a more rational approach to decision making, the second self-perception is the more useful.

Internal evaluation is the distinguishing characteristic of Wildavsky's (1972) self-evaluating organization. The internal evaluator is Wildavsky's (1972, p. 510) "evaluative man": "The ideal member of the self-evaluating organization is best conceived as a person committed to certain modes of problem solving. He believes in clarifying goals, relating them to different mechanisms of achievement, creating models (sometimes quantitative) of the relationships between inputs and outputs, seeking the best available combination. His concern is not that the organization should survive or that any specific objective be enthroned or that any particular clientele be served. Evaluative man cares that interesting problems are selected and that maximum intelligence be applied toward their solution. To evaluative man, the organization doesn't matter unless it meets social needs. Procedures don't matter unless they facilitate the accomplishment of objectives encompassing these needs. Efficiency is beside the point if the objective being achieved at lowest cost is inappropriate. Getting political support doesn't mean that the programs devised to fulfill objectives are good; it just means they had more votes than the others. Both objectives and resources, says evaluative man, must be continuously modified to achieve the optimal response to social need."

Wildavsky's definition of evaluative man does not mention any particular discipline or preparation for the internal evaluator. The defining criterion focuses on problem solving and model building from a rigorously analytical perspective. In fact, it is our view that model building is one of the evaluator's main tasks. The evaluator has the primary function of asking what-if questions and of conducting both hypothetical experiments and real studies to answer them, so that the function has real utility for planning and decision making. The evaluator is a professional challenger of assumptions, values, and constraints who helps to develop alternatives. The one point at which we differ from Wildavsky involves the loyalty of the internal evaluator. While evaluators do care about interesting problems and how they are solved, the internal evaluators with whom we have worked in mental health are also deeply committed to efforts to help the mentally ill and developmentally disabled. Internal evaluators need this kind of commitment to maintain themselves through the years required to build a self-evaluating organization and to have some real and sustained impact on the management, planning, and organization of mental health services.

Evaluation was not cut from whole cloth by social scientists. It is an integral part of the management process. Good management involves the continuous monitoring of operations and the application of control mechanisms to bring operations into line with prescribed objectives. The difference between evaluative man and administrative or management man lies in their repective orientations or in the point at which each begins. As Wildavsky notes, evaluative man is motivated to apply maximum intelligence to the problems facing the organization, whereas administrative or management man is motivated by organizational survival and the pragmatism required for shorter-term success. Nonetheless, the ethical manager, like the evaluator, is a skeptic who periodically and critically looks at the rationale for the organization and its programs. Thus, evaluators and managers begin with different motivations, objectives, and working styles. A continuing primary task for the evaluator is to bridge the gap between the two and ultimately to bring about a marriage between them.

Roles and Skills. Traditionally, evaluation has been viewed as applied research aimed at discovering the relationship between program inputs and outputs. Additionally, evaluators in mental health services have frequently described their role as that of social researchers who desire to increase the fund of knowledge about mental health and mental illness.

As already noted, we view the internal evaluator's basic role as that of management decision-support specialist. In this role, the evaluator aims to enhance the ability of managers to perform their planning and control functions. As a result, the evaluator must have both technical and analytical skills and interpersonal and organizational skills. Above all, he or she must be able to recognize where and when these analytical, management-support skills can best be applied. The internal evaluator needs to keep in mind that the evaluator's function is to support the decision-making process, not to usurp it.

Internal evaluators also have opportunities to function in a number of other roles, including planner, operations researcher, manager, organizational development consultant, management trainer and consultant, and data-processing or information specialist. In the context of helping to improve management by supporting a more rational decision-making process, the internal evaluator must be prepared to play any and all these roles at one time or another. The internal evaluator will play some of these roles more often than others. Data-processing or information specialist, planner, and operations researcher will be the most frequent, in that order.

By any definition, evaluation is an interdisciplinary venture. Given the scope of management tasks and the variety of roles that the internal evaluator can be called on to play, the internal evaluator must develop a wide range of skills and have the ability to acquire new skills quickly.

Data Processing or Management Information Specialist. The currency of the evaluator's trade is information. The responsibility for designing, developing, implementing, and maintaining data systems is a fundamental concern of internal evaluators. Successful implementation of managment information systems requires internal evaluators to analyze systems requirements, design systems, develop procedures to ensure the validity and integrity of data, and design systems outputs that maximize use of information in management's decision making. The skills just specified are required by both manual and computerized information systems. The computerized environment requires additional knowledge, including knowledge of automated systems design, computer operations, computer languages, and the capabilities of hardware and software.

Data Analysis, Interpretation, and Presentation. The decision-support philosophy requires someone in the organization to have the ability to obtain, aggregate, analyze, and interpret data. This individual must also be able to present data in a way that is useful to managers. The skill set required here is basically the same set that people trained as social researchers possess: statistics, the ability to manipulate computer packages that perform data analysis, writing ability, and the communication skills needed to package data in such a way as to ensure their impact on the intended audience.

Communication and Interpersonal Skills. Much of what internal evaluators are able to accomplish in an organization is contingent on their communication and interpersonal skills. By communications skills, we mean more than just the ability to ensure that the message which the listener receives is the message that the speaker intended. Because the role of the internal evaluator often includes management consultation, he or she also needs to be able to listen. Active listening enables the evaluator to comprehend the speaker's message and convey to the speaker that the message has been received. The evaluator must also be able to take diverse and often vaguely formulated ideas, synthesize them into a clear, concise whole, and feed this formation back to the organization as a group product. He or she must be able to communicate at a level that is neither pedantic nor condescending.

Conflict resolution skills are also important. Much of what evaluators are called on to do requires them to function as organizational change agents. This role is crucial in the early stages of designing and implementing data systems. The process of constructing data systems (and the issue of building accountability systems in general) often places the evaluator in the position of having to convince clinical people that management systems are needed. The process demands a good set of conflict resolution skills, since there are often inherent conflicts between what is desirable from the clincial perspective and from the management perspective. The specific skills involved are those required to help people overcome their resistance to change. The ability to separate real issues that must be addressed from resistance issues is crucial. The change process requires the ability to get people to state their concerns, then to devise methods that can satisfy these concerns without sacrificing the original objective.

Ability to Adopt the Manager's Perspective. In order to prompt a decision-support orientation and a data-based approach to management decision making, evaluators must be cognizant of and able to empathize with management's role in the organization. Thus, the evaluator must be able to function as a manager. We believe that the ability to empathize with managers is sharpest among evaluators who have walked a mile in the manager's shoes. That is, the experience of balancing political, organizational, and clinical realities with empirical data and good management practice is essential for the evaluator. The skills derived from such experience enable the evaluator to design more effective information systems and to conduct studies that will actually be used in decision making.

Ability to Function as Planner and Operations Researcher. The planner role is associated with individuals in the organization who have responsibility for anticipating future events that will have consequences for the organization so that planning can start in the present. The skills required to perform this function include the ability to synthesize a great quantity of fairly ambiguous data about the future environment. It includes the organizational skills required to influence decision making so that the organization periodically questions its assumptions and critically views its mission and procedures. The ability to do what-if contingency thinking is another requisite skill.

The same what-if skills are required to determine the probable impact of a decision on the organization's current functioning. These skills are called for when the evaluation specialist is playing the decision-support role of anticipating the consequences of alternative solutions to a problem. The skills required are those needed to perform modeling and simulations that enable the probable outcomes of alternative management actions to be projected. Being able to perform this function on one's feet is critical, since the injection of evaluation into decision making requires the evaluator to identify on the spot where and how specific information can be brought to bear on a given problem. Such modeling and simulation can be highly abstract, conceptual, and qualitative, or it can involve the use of formal mathematical techniques and empirical research. Facility with spreadsheet programs, forecasting procedures, linear programming techniques, and other quantitative techniques is very useful.

Organizational Development Consultant and Management Trainer. More often than not, it is the role of the aspiring internal evaluator to promulgate the philosophy of a data-based approach to decision making and to educate the organization about the evaluator's role as a decision-support person. We conceptualize these tasks as organizational development tasks. The skills required include most described in the section on communication and interpersonal skills as well as the ability to proselytize new converts.

Once people have been converted to the data-based approach, the evaluator may be called on to assume responsibility for training managers to use data to improve their decision making. This process involves the skills necessary to sensitize managers to their responsibilities and to provide them with the skills that they need in order to incorporate data into their decision making. New managers need help in conceptualizing their responsibilities as decision makers. Managers have difficulty in responding to such questions as, What information would you like to have? They have much less trouble specifying the information that would be useful in making specific decisions (Sherman, 1981). The process of getting managers to conceptualize their responsibilities as decision makers requires skills in management training.

Bringing About a Marriage Between Evaluation and Management

The first task in bring about a marriage between evaluation and management is to adopt the notion that such a marriage is not only feasible but also desirable. The preceding sections of this chapter have spelled out our view that evaluation and management are intimately related and that they need to be formally wed in the organization.

Management is a continuous process of assessing and reassessing opportunities and operations, of controlling and correcting, and of moving in new directions as conditions allow or dictate. Support for management in these processes must thus be continuous (that is, formative), not discrete (that is, summative). In all instances, management is driven by the discipline of the cost-benefit ratio. Management does not wait for the final report to be completed before making a change if the costs outrun the benefits and if there is no reason to believe that the trend will be reversed.

It is important for the internal evaluator to keep the distinction between evaluation and evaluation research in mind. Although Suchman (1967) has made this distinction clearly and explicitly, it often becomes muddled. For instance, in its widely disseminated handbook on evaluation techniques for community mental health agencies (Hagedorn and others, 1976), the National Institute of Mental Health defined evaluation as a type of applied research. The distinction between evaluation and research is not academic, because it affects the evaluator's role and how he or she operates. The internal evaluator cannot survive or function effectively as an applied researcher. As Wildavsky (1972) points out, the internal evaluator is a problem solver, an adjunct and support to the planning and management process, a change agent who chooses to work inside the system in order to improve an organization's efficiency and effectiveness and increase the ability to change from within.

The same distinction lies at the heart of the difference between formative and summative evaluation. Formative evaluation appraises an ongoing program in order to modify and improve it. The activities of formative evaluation include analysis of management strategies and of the interactions among persons involved in the program as they affect the program's operation (ERS Standards Committee, 1982). Summative evaluation is a type of applied research and requires many of the same conditions as any sound social experiment: controls over as many of the nonexperimental factors as possible, explicit and operational outcome criteria, and explicit estimates of baseline levels of dependable variables before the program being evaluated is applied. Formative evaluation is the primary and defining mode of the internal evaluator. Although one may attempt to apply controls, get definitive pretreatment measures, and otherwise do very rigorous research, all these efforts are secondary to the task of improving the efficiency and effectiveness of ongoing programs. The managers are constantly tinkering with programs, and management strategies—both programmatic and political—are dynamic. The evaluator is inevitably caught up in the interaction and limitations of particular managers, clinicians, and other actors. The process poses a challenge to the evaluator's skill as a scientist and manager and to the evaluator's personal and ethical position. This is the price that the internal evaluator has to pay to bring his discipline to bear on a problem, and it is the challenge to the integrity of that discipline.

Summative evalution can actually be a trap, politically and organizationally as well as conceptually. Summative evaluation assumes that both the environment and the program are stable for the period of the evaluation. The internal evaluator knows that this assumption is incorrect for any significant period of time and for any program of significant breadth. Management's task is to adjust, adjust, and adjust in the light of feedback on the program and the environment. Most of this feedback is informal, and it comes from many sources. Live agencies are simply not laboratories. Because of the disciplinary as well as the political constraints, summative evaluation almost always focuses on a short time period. In this temporal framework, its commitments cannot be to real organizational learning, development, and change. The summative evaluator is an outsider to the organization being evaluated. Outsiders have little leverage for change. Moreover, it is very unlikely that they will have the time required to bring about fundamental changes in the organization that delivers the program that they are evaluating.

It is not the purpose of this section to go into detail about specific activities but rather to identify the issues and types of priority activities in which the evaluator should engage. In the next section, we will discuss the markers of success that give evaluators a standard by which to evaluate the success of their strategy and tactics.

Raising Managers' Consciousness. There has been considerable discussion in the literature and among evaluators about how to make managers, clinicians, board members, and others better consumers of evaluation. From our perspective, this effort is misplaced. For evaluation to be useful and to be used, the managers have to accept responsibility for owning and defining the evaluation function. Consumption is much too passive an activity if evaluation is to play a major role in an organization's decision making. As long as managers view themselves only as consumers of evaluation, not as active participants, evaluation will remain marginal.

The objective is not to turn managers into evaluation technicians. There is no question that a whole variety of specialists is required to carry out the complete range of evaluative activities. As the discussion earlier in this chapter makes clear, no one person possesses all the skills needed to meet a mental health center's total evaluation requirements. Thus, the objective is to meld management support staff into a functional team and to help managers to become better managers. For that reason, the objective is to bring managers to the point where they ask the evaluative questions when they are considering policy and program alternatives. The questions cannot be posed as afterthoughts. Managers need to ask how they will know that something is working or how they can find out what, if any, changes need to be made while they are still at the planning stage.

Consider how an ideal discussion of program alternatives proceeds. As one considers an alternative, one sets up a conceptual model of how it will operate; then, varying the inputs and the model's parameters, one considers the range of outcomes. In situations where the knowledge of how a system works is limited — this is often the case in mental health services — this exercise is highly conceptual, and a number of assumptions have to be made. Where the evaluator and others can bring some knowledge and data to bear on the situation, the uncertainty is reduced to some degree, and the quality of the decision improves as a result. As already noted, initiating and leading such discussion is one of the evaluator's key roles. Two of the evaluator's most important skills are first, to construct models over a broad range of levels of abstraction; then, to simulate, conceptually or with data, a system's behavior so as to predict data to bear on a situation to help reduce uncertainty. At the same time, the evaluator helps to evaluate the usefulness of the resulting information as well as the costs and benefits of getting additional information. When this process is followed consciously, it is effective in raising the consciousness of both managers and evaluators. One cannot expect managers to request data and complex special studies until they can think systemically and make decisions in this relatively sophisticated way.

Being Actively Aware of the Costs of Evaluation. Every management decision is made in a context that weighs financial, programmatic, and political costs against benefits. Like any other activity, evaluation needs to be put to the cost benefit test. Evaluation and planning are not beneficial in any obvious way, except perhaps for those who make a living conducting evaluations. Thus, the evaluator needs to take the lead in assessing where additional information may significantly reduce uncertainty or risk and weigh the expected benefit of this reduction against the probable cost of obtaining the needed information.

Two comments are in order here. First, the evaluator actively and continuously considers the costs of evaluating the organization's activities. Evaluation proceeds only as long as the benefits exceed the costs. Costs include not only monetary outlays but opportunity costs and political costs as well. Second, the evaluator does not make any assumptions about the inherent value of evaluation for the organization. This stance is essential if the evaluator is to build personal and professional credibility. It also serves as a model for all management decisions.

Negotiating and Merging Styles. In our discussion of the functions of evaluating and management, we touched on the assumptions and operating styles typical of both roles. To create an effective working relationship and beyond that to bring about a marriage between the two requires a merging of the two styles. The beginning point for such a merger is the attitude that neither style of operation and decision making is inherently better than the other. Both styles contribute to the success of the organization, and different criteria and different time frames must be used to assess their performance. Movement toward some kind of synthesis or merger contributes to both the quality of management and to evaluation; thus, it promotes the functioning of the organization as a whole.

The hybrid style has a short reaction time, and it is proactive in anticipating both future changes in the environment and the secondary consequences of actions. The hybrid style is highly eclectic about the sources of data used for decision making. It is critical of informal sources, and it values systematized ways of organizing and judging the quality and value of information. It values empirical research for its scientific rigor and it uses empirical research because it addresses real decision problems in a timely and politically sensitive way. Reports are kept short. Often, reports are oral, and formats are varied for different audiences and settings. Introducing the hybrid style involves changes in both management and evaluation. These changes result from mutual education and accommodation aimed at improving the organization's effectiveness and efficiency by improving its planning and decision making.

One important aspect of merging the two styles involves moving the organization away from informal approaches to evaluation (casual observation, implicit objectives, intuitive norms, subjective judgments) toward the formal approaches to evaluation (Caro, 1971). Perhaps the most powerful lever in bringing about such movement is the development of formal management information systems. From a developmental perspective, the management information system has always been viewed as a primary (if not the primary) step in developing evaluation capacity (McIntyre and others, 1977). Careful management of the data system as it evolves is very important in the growth of management and its movement toward the data-based style of decision making. A management information system that has real utility for the organization must mirror the organization both in its structure and in the models that it uses for decision making. It takes painstaking work with management to make these structures and models formal and explicit. The status quo of power and resource distribution within the organization as well as many other aspects of its modus operandi exist because they are implicit, unexamined, and unstated. Making the objectives of the organization and each of its subcomponents explicit can be extremely threatening. To encompass these objectives in a formal management information system means defining these aspects in operational terms so they can be observed, measured, analyzed, and reported.

Of course, in such a field as mental health, the ideal information system that specifies everything in exhaustive and mutually exclusive categories and that neatly fits everything into a model relating system inputs to outputs is a dream unrelated to reality. Most human services are still beyond our modeling capabilities. Indeed, defining many of the critical terms — *illness, functioning, personal resources* — in any reliable and valid way is beyond our capabilities. The evaluator cannot promise any easy answers or offer any tricks for planning and decision making. Thus, much of any management information system is experiments — hypotheses about how things work, hypotheses about the information that would be helpful in managing these things, and a formal way of testing these hypotheses.

The evaluator invites managers to adopt a more experimental — that is, a skeptical, error-embracing, change-oriented — approach toward management. But, management is an activity that modally avoids uncertainty, denies error, and minimizes risk. Management seeks to maintain the status quo — the dynamic balance among elements of the political coalition on which survival depends. That there will be resistance is obvious. One of the levers is the fact that sophisticated data and plans are useful weapons in the political arena. The requirements of funders have become more complicated in recent years, if no more sophisticated, and the dynamics of human service organizations have become so complex that they outstrip traditional management approaches to decision making in all areas.

This point is important. We are advocating not just a merger of decision-making syles but a fundamental change in the operating style of managers and organizations. Wildavsky (1972) adopts this position when he notes that self-evaluation means a continuous process of change for the organization. Change is threatening. It requires, he notes, new modes of decision making, new political arrangements, and creation of new levels of trust among organizations, communities, and funders. The futurist Donald Michael (1973) makes many of the same points. He, too, speaks of the need for change in organizational decision making: to live with uncertainty, embrace error, seek and accept ethical responsibility, evaluate current operations in the light of future needs, live with role stress, and be open to changes in commitments and directions as suggested by conjectured pictures of the future and by evaluation of ongoing activities. Such changes are needed if the organization is to survive and to serve us well in an increasingly unstable and rapidly changing environment. The factors that Michael specifies as central to this process coincide with the values of the internal evaluator as we view them. The stakes are important, as is the role of the evaluator.

Organizing the Internal Evaluation Team. In order both to establish a strategic role for evaluation in the organization and to bring about a marriage between evaluation and management, the internal evaluator has to consider the composition and organization of the internal evaluation team, its location in the organization, and the range of functions that it performs.

To occupy a viable and important niche in the organization, evaluation cannot often live by evaluation alone. To secure a place in the planning and decision-making processes, the evaluator may have to assume some other important organizational tasks, such as training, budgeting, finance, planning, and quality assurance. The actual tasks that the evaluator assumes will depend on his or her personal skills and on the organization's strucure and needs. The important thing to keep in mind is that evaluation alone will not suffice to give the evaluator access to all the settings in which important decisions are made. Evaluation alone will not place the evaluation in a position to know critical information. That kind of position most often has to be earned bit by bit. It is seldom bestowed on initiates. It also means that the evaluator has continually to acquire new competences and that the evaluator needs the personal flexibility and skill required to mesh new tasks with his or her role.

The issue of power and influence is an important one. Some minimal level of power or influence is necessary just to conduct the day-to-day business of evaluation. To promote organizational change and to see that one's products receive serious consideration in decision making takes more than just professional or technical credibility. Access to the top is critical. It is not, however, sufficient. First of all, as a member of "support" staff, the evaluator has no real power in the organization. Evaluators will seldom, if ever, be in a position to compel compliance on their word or personal power alone. What the evaluator can have is influence. This influence, which is a kind of power, can come from two sources. It can be referent power from one's peers — middle managers—or it can be delegated power from the executive director. Although the evaluator should seek to be attached to the office of the executive (with no intervening supervisor or manager), reliance on delegated power alone is isolating. It is essential to seek referent power from one's peers and the other managers. Such power is gained by providing useful services. The discussion of the functions of evaluation in the organization earlier in this chapter gives several examples. This is another reason why the evaluator must be prepared to build a diverse portfolio. The two types of power feed each other. To concentrate on one is ultimately to diminish the other and alienate those who can grant or delegate it.

The strategy just outlined is not without its dangers. Both the playing of multiple and important roles and intimate access to top decision makers can be a two-edged sword, especially if the evaluator assumes responsibility for such things as budgeting, accounting, and other aspects of the financial and resource allocation mechanism. Being close to the money can feed the paranoid feelings of clinical staff and lead staff to assume that there are tacit, not well understood connections between evaluation and the allocation of funds. At the least, this may put middle managers and clinicians on the defensive. At the most, it can sabotage legitimate and supported evaluation projects. The realities of the situation need to be made explicit. Any fears or resentment need to be worked through. It is very important to keep the ethics of the evaluator's role in mind. A cardinal rule is that, as "support" staff, the evaluator must not put himself or herself in the position of making executive or line decisions.

Markers of Success

Numerous specific strategies and tactics can be used to bring about the marriage of evaluation and management advocated here. Some of these strategies are implicit in the preceding discussion. The exact strategy that an evaluator chooses will depend on his or her particular skills, the needs of the organization, and the politics of the immediate setting. Thus, instead of spending time on these specifics, we feel that it will be most useful to discuss eight markers of successful internal evaluation. This discussion is based on our combined experience of fifteen years in community mental health and on our discussions with other evaluators. Obviously, our choice of markers is contingent on our definition of evaluation and the evaluator's role in a community mental health center. Because the evalutor's purpose in our view is to improve the organization's planning and management by applying information and rational models to decision making, our markers relate to changes in the organization, in the way in which it operates, and in the way in which it uses the evaluator's skills. Many of these markers are simple indicators of the perceived usefulness of evaluation's products, while others reflect more profound changes in people's thinking and ways of making decisions.

First, when you are late with the outputs from the data system, people call to find out where they are. This is a simple market measure of the value of one of the evaluator's products. At times, it may even be instructive to hold up or not run certain reports to test the effect of their absence. Many reports and analyses are useful for a while but then lose their utility.

Second, you have to work hard to prioritize the requests that you receive. This indicates that evaluation products and your input are valued: The number and range of organizational units that seek your input are important, because they indicate importance of evaluation in the organization's program and interprogram decision making. Finally, the number of levels of the organization that are conducting evaluation on their own is an indicator both of the extent to which you have been successful in making evaluation important and of the extent to which you have helped to empower people at the lower levels of the organization, giving them information and tools with which to become involved in planning and decision making.

Third, you and the people who work for you assume increasingly important roles in the organization, particularly in areas once considered clinical reserves, such as the quality assurance committee. The extensiveness and the importance of evaluation staff involvement in quality assurance is a critical aspect of this indicator. The evaluator should and almost assuredly will be involved in retrospective reviews that analyze data from the center's data system and perhaps clinical records as well. The successful evaluator will have a regular seat on the committee, and he or she will also be involved in standards development, concurrent reviews, and all other aspects of the center's quality assurance program. The successful evaluator's skills and systemic perspective will have considerable impact on the choice of topics for quality assurance attention and on the committee's orientation to its mission. The evaluator really knows that he or she has arrived when he or she or a member of evaluation staff is chosen by clinical colleagues to chair the quality assurance committee, since it is such a critical setting for evaluation and for the success of evaluation within the organization. This committee will almost always represent a cross section of disciplines, programs, and levels within the organization. It is therefore an important setting for educating the organization about evaluation and about how it can be used, not just in itself, but in its models and systemic ways of thinking as well. The committee is an important channel for reaching and ultimately helping to empower the nonmanagerial levels of the organization. Perhaps most important, it is the one place where clinical (nonmanagement) staff can see and understand the short-term utility of evaluation.

Fourth, clinicians stop arguing that you cannot possibly understand them or their situation because you are not a clinician. This argument and the frustration that it often reflects are directed not only at evaluators but at all management people at one time or another. Each level of the organization feels that the levels above and below it are out of touch with what is really going on. On one level, cessation (or a decrease in frequency) of this kind of remark will reflect a gain in personal acceptance and credibility within the organization. It shows that one has gained trust and demonstrated that one's efforts are useful. On another level, it reflects a change in the way in which other people think about what they do and how they do it. A manager or supervisor who has begun to think more systematically and systemically is less likely to take this position. A clinician who has begun to think beyond the individual client to populations of clients and to the impact of programs on these populations will make fewer remarks of this nature. These people have begun to think like managers.

Fifth, you are increasingly called in because of your impartiality. Your systems perspective and your ability to listen, synthesize, and give feedback are valued.

Sixth, knowledge is increasingly related to power in the organization. By *power*, we mean the ability to be meaningfully involved in and influential in decision making. That is, decision-making processes are becoming increasingly rational and less confined by tradition, personal power, bureaucratic hierarchy, and vested interests. Data are available on most aspects of the agency's operation. People are struggling to formulate models of what they do that show how inputs are related to outputs. Decision makers are taking a more systemic view and are conscious that decisions involve trade-offs. As the data-based decision-making mode begins to predominate, the number of people involved in decision making increases. If the evaluator is successful, there is no monopoly on information at the top of the organization. People at all levels of the organization have access to the information that they need in order to become involved in planning and decision making.

Seventh, the issue of ethics and organizational conscience has moved outward and downward. The successful evaluator is not in the untenable position of trying to be the organization's conscience or the sole focus of accountability.

Eighth, the organization has increased its ability to change. This may be the single best indicator for the self-evaluating organization. Such an organization continuously monitors its performance, then uses the feeddback to improve its efficiency and effectiveness while looking critically at the rationale for continuance of its programs. This does not mean that the organization turns itself inside out every year, but it does mean that at any point in time the organization is likely to be critically reviewing some aspect of its operations and that change follows from analysis. The time required for analysis, discussion, and implementation will decrease over time.

Ethical Issues

Ethical issues for evalutors have been discussed extensively over the past several years. These discussions led to development of elaborate standards by the Evaluation Research Society (Rossi, 1982) and others. These standards and much of the literature are concerned with what we call *task-specific* ethics — ethical standards for data collection, analysis, interpretation, presentation, and so forth. Our concern here is to offer some observations on ethical issues for the role as a whole. Much of the debate is based on oversimplified models of the role of the practicing internal evaluator. Evaluators who place themselves in the position of advocates who can produce any result that a client needs or who — at the other end of the continuum — are accountability specialists who look for error, wrongdoing, or incompetence in funded agencies do not last long as evaluators. The requirements of the role, the enterprise, and the social settings in which evaluation is conducted are too complex for such one-dimensional behavior. From our point of view, the main issues in the ethics debate center on the myth of objectivity, the academic bias toward summative evaluation, and naive assumptions about the evaluator's ability to manipulate and to be manipulated.

One of the touchstones of the scientist and therefore of the evaluator as scientist is the evaluator's objectivity. Objectivity is usually defined as the ability to stand outside a situation and to render an unbiased observation of what has "really" occurred in a particular situation. This objectivity is a myth that most thoughtful scientists reject. A more viable definition of objectivity is provided by C. West Churchman (1968, p. 86): "Instead of the silly and empty claim that an observation is objective if it resides in the brain of an unbiased observer, one should say that an observation is objective if it is the creation of many inquirers with many different points of view. What people [organizations] are really like is what people with the strongest of inquiring motivations will perceive themselves to be like. The 'verification' of scientific findings resides in the creative spirit of human inquiry carried to its maximum potential."

For us, Churchman's definition of objectivity is the touchstone of the internal evaluator. In our discussion of markers of success, we noted that ownership of evaluation by management and other members of the organization and empowerment of people through active participation in and open access to information derived from evaluative activities were two of the evaluator's main strategic objectives. Churchman's definition of objectivity clearly makes these scientific and ethical objectives as well. Part of the ethical dilemma for the evaluator is that both he or she and the evaluation function are often (more often in the literature and academic debate than in the field) cast in the role of organizational conscience as the primary conduit of accountability to funders, the governing board, and the community. It is relatively easy to adopt this position, particularly if one is young and idealistic. However, the position is untenable for both the evaluator and the organization. In the short term, it may remove some difficult issues to a safe and controllable distance and appear to let some people off the hook. In the long run, it is both bad management and bad politics, and for that reason it has little real survival value.

Another concern is raised by the questions of whether the evaluator is manipulated by managers and policy makers and whether the evaluator manipulates study or research results so as to promote his or her own preferences. The first issue is not so much one of manipulation as of the evaluator's being coopted by the narrow and partisan interests of organization managers, policy makers, or both. Consultants are often seen in the same light as internal evaluators — hired guns whose job is to act as advocates and shoot it out with funders and competing programs. In contrast, the academically based evaluator is viewed as objective and unbaised. On the purely individual level, manipulation can be a problem. Ethical resolution of such a problem depends on the personal integrity of the individual evaluator. In practice, however, the problem is neither as common nor as serious as it may appear. First, evaluation is a social enterprise requiring the cooperation of a great many people, whose interests in the outcome are not identical. Manipulation requires a conspiracy that for all practical purposes is impossible to achieve. Second, most people are fairly sophisticated when it comes to evaluation, and they are not easily fooled. Besides, if the theory and the data analysis used in the evaluation are so arcane that the users do not understand them or if the results are so counterintuitive or so skewed that no one believes them, then the evaluation will carry little weight in the final decision. Third, as Patton (1978) has so amply demonstrated, formal evaluations are only one factor in evaluation of programs and in the decisions that affect programs and policies. The bottom line is that manipulation of and by evaluators, particularly of and by internal evaluators who work in an organization that is part of a service system, is not easy, and it is extremely counterproductive in the long run. In this regard, we can formulate one rule for current and would-be internal evaluators: No matter how great the temptation, the evaluator should not encroach on the program manager's turf. Maintaining one's status as a support to management decision making and planning is essential for both credible and ethical evaluation. Encroachment is often the beginning of the whole manipulation syndrome.

The internal evaluator, who focuses on formative evaluation, faces a value conflict less often than the external evaluator, who makes summative judgments about programs. Windle and Neigher (1978, p. 105) make this point in their advice to trapped evaluators: "Ethical problems are most likely when program evaluation is used for 'summative' evaluation. . . this type of information is useful mainly for decisions about program support made by persons outside the program being evaluated. The program's posture regarding such research is clearly biased, leaving the evaluator in a vulnerable position. 'Formative evaluation,' however, looks into the program conditions associated with relatively good performance and is thus a basis for making choices and changes within a program to improve its effectiveness or efficiency. Such evaluation can be fairly neutral for the overall program manager, since the program managers will usually not care greatly which of several options is followed as long as better results are obtained for the program as a whole. Such neutrality will not, of course, be true of managers of program components which are compared, but the program evaluator who works for the manager of the total program can maintain a relatively unbiased position."

In formulating their prescription for avoiding one ethical dilemma, Windle and Neigher identify another dilemma, for which they do not supply an antidote. The issue of different levels or components of the organization or service system returns us to the question, When is the evaluator being summative, and when is the evaluator being formative? When is the evaluator working in a participative environment, and when is the evaluator working in a "political," manipulated or manipulating environment? This ethical issue is dealt with in our prescription for bringing about a marriage between management and evaluation. One of the markers of success is the involvement of a wide spectrum of organization members in the evaluation enterprise. Creation of a community of interest in the refinement and growth of the capabilities of a human service enterprise both among organization members and in the larger system of funders and policy makers is one of the evaluator's tasks. Wildavsky (1972, p. 519) notes that "the acceptance of evaluation requires a community of men who share values." He also concludes (p. 520) that "evaluation need not create agreement; evaluation may presuppose agreement." If we recall Churchman's definition of objectivity, we see that Wildavsky's view makes involvement an ethical as well as a practical political matter for the evaluator. The promotion of a community of interest and of a systemic view of the agency and its subcomponents serves the ethical function of involving many others in the ownership of a larger set of responsibilities and accountability. It removes the evaluator from the uncomfortable position of trying to be or of being seen as the organization's conscience.

Growth and Future Trends in Internal Evaluation

The future for internal evaluators is promising. A combination of political and economic factors ensures that internal evaluation will continue to grow. The same factors will probably result in a reduction in the number of mental health evaluators who have not made the shift to the decision-support role.

Federal Legislation. One of the factors that has served to reduce the number of externally oriented mental health evaluators is revocation of P.L. 94–63, the federal law that mandated evaluation. Regulations that accompanied the law may have been responsible for the tremendous growth of evaluation positions in mental health. The same regulations may have misled many of those who were employed under their aegis into believing that their positions were safely sheltered and that the existence of the law would be sufficient to ensure that their efforts would have an impact on the mental health agencies that employed them. At the point when the federal government decided to evaluate the impact of evaluation on mental health centers, it was difficult to find centers where impact could be documented (Flaherty, 1978). This lack of impact, or at least the absence of documentation, was responsible in part for the removal of the federal mandate for evaluation and for the substitution of externally dictated performance-based measures of the extent to which system goals and objectives were being met. As a result, performance measurement systems have become the mechanism by which the mental health service delivery system is evaluated and managed. The first federal effort to implement performance measurement systems—the Operations Management System (Sherman and others, 1981)—was short-lived. Replacement of the Mental Health Systems Act by the Omnibus Budget Reconciliation Act in 1982 shifted the responsibility for accountability to the states. Many states have chosen to implement performance measurement as the system of accountability.

The net effect of these legislative changes is that centers have been able to eliminate evaluator and support staff jobs. For the most part, the jobs that were eliminated seem to have been occupied by evaluators who never quite managed to abandon the agency conscience role advocated by the evaluation literature or to adopt the decision-support role advocated in this chapter. Thus, the people who were eliminated were never integrated into the management structure of their agencies. Many members of a third class of evaluators are still in residence in mental health settings. These individuals cannot decide whether to be full-fledged decision-support types or simple data-processing managers. The distinction between the two roles is whether the role includes consulting with management about the interpretation and implications of data for management decision making or whether it is limited to producing numbers.

Resource Reduction. The second set of factors is occurring in the economic sector. The scarcity of resources for mental health programs has increased the pressure to maximize utilization of resources. In many organizations, this pressure to "optimize" the use of scarce resources represents the first real push to manage and control operations proactively. As a result, the support for and interest in management is increasing in human service organizations. Any interest in increasing the level of proactive management carries with it an increased role for the decision-support function. Agencies that implemented data systems to satisfy external reporting requirements now find themselves having to rebuild those systems so as to support the internal decision making that "optimization" requires. It is paradoxical that a reduction in funding should promote the growth of internal evaluation.

It may be important that management experts have long decried the state of management in nonprofit organizations (Borst and Montana, 1977) and that evaluation experts have identified management's lack of sophistication as a major factor in the failure to use program evaluation (Attkisson and others, 1978). The forced "optimization" can be an opportunity for both management and evaluation. Improvement in both depends on bringing about a workable marriage between the two.

Prepaid Health Care. We are entering an era of proscription. While the regulations that accompanied the federal mental health legislation were full of admonitions about the structure and process necessary for a functional mental health operation, federal funding was never tied to performance. There were at least two reasons for this state of affairs. The first was that one of the superordinate goals of the federal program was to build a mental health service delivery system. One gross indicator of success in achieving that goal was the number of centers in operation. The second reason was related to the difficulty of managing such an operation at the federal level. In any case, there were few negative contingencies for centers that failed to meet expectations. The contingency management of such programs is just now coming of age, oiled by the scarcity of resources and by the delegation of responsibility to the states. Thus, we now see states specifying reasonable unit costs and identifying the clients whose treatment can be covered by state monies. Competition has increased and with it the recognition that attempts to preserve the monopolies created by the federal catchment area concept do not represent the most cost-effective way of delivering mental health services. The system is fluid, and competition for clients who can pay for services has increased. Better-managed operations will obtain larger amounts of the available resources. The quality of decision making will improve as the stakes continue to rise. The trends in mental health will parallel those in medicine, where competition for patients has also increased. Diagnostic-related groupings, prospective payment organizations, and increasing use of prepaid health maintenance plans for the provision of care to the indigent are appearing on the horizon. Mental health organizations that have personnel who can manage effectively will reap larger shares both of this market and of the market consisting of individuals who have private insurance. Increasingly, survival will depend on the ability to deliver quality care at the lowest possible cost. The incentive for providing the least-restrictive care will also be the least costly to funders.

Outcome Data. Demands for demonstrable mental health treatment outcomes can be expected to escalate. States are just beginning to demand that efficacy of treatment be demonstrated as a condition of funding. The role of evaluators in the design of outcome assessment systems and in the interpretation of outcome data will grow, if for no other reason than the role that quality outcome data have as a marketing device.

System Accommodations and Professional Role Identification. Future historians of social science may view internal evaluation in particular and evaluation in general as a social movement within the human services. As with all social movements, the parts of the movement that are successful in promulgating change will be assimilated into the establishment that they set out to reform. As a consequence of this assimilation, the movement experiences a

44

need to develop a different set of issues as the reason for its existence. We believe that an increasing proportion of the internal evaluator's work will involve the introduction of management science techniques into the mental health environment. This movement will increase the chasm between academic-based notions of evaluation and internal evaluator's functions. The time may well come when internal evaluators will align more comfortably, philosophically, and professionally with the field of operations research than with the field of evaluation. This trend is already evident among internal evaluators who are seeking a more concordant peer group. It may also be the harbinger of new reform movements that may increasingly rely on the sophisticated use of information and models to make their case.

References

Anthony, R., Dearden, J., and Vancil, R. *Management Control Systems.* Homewood, Ill.: Irwin, 1965.

Anthony, R., and Herzlinger, R. *Management Control in Nonprofit Organizations.* Homewood, Ill.: Irwin, 1980.

Attkisson, C., Brown, T., and Hargreaves, W. "Roles and Functions of Evaluation in Human Service Programs." In C. Attkisson, W. Hargreaves, M. Horowitz, and J. Sorensen (Eds.), *Evaluation of Human Service Programs.* New York: Academic Press, 1978.

Borst, D., and Montana, P. (Eds.). *Managing Nonprofit Organizations.* New York: AMACOM, 1977.

Caro, F. "Evaluation Research: An Overview." In F. Caro (Ed.), *Readings in Evaluation Research.* New York: Russell Sage Foundation, 1971.

Churchman, C. W. *Challenge to Reason.* New York: McGraw-Hill, 1968.

ERS Standards Committee. "Evaluation Research Society Standards for Program Evaluation." In P. H. Rossi (Ed.), *Standards for Evaluation Practice.* New Directions for Program Evaluation, no. 15. San Francisco: Jossey-Bass, 1982.

Flaherty, E. "A Review of Community Mental Health Centers' Evaluation Activities." Unpublished report, Philadelphia Health Management Corporation, 1978.

Hagedorn, H., Beck, K., Neubert, S., and Werlin, S. *A Working Manual of Simple Program Evaluation Techniques for Community Mental Health Centers.* DHEW Publication No. ADM 79-404. Washington, D.C.: U.S. Government Printing Office, 1976.

McIntyre, M., Attkisson, C., and Keller, T. "Components of Program Evaluation Capability in Community Mental Health Centers." In W. Hargreaves, C. Attkisson, and J. Sorensen (Eds.), *Resource Materials for Community Mental Health Program Evaluation.* (2nd ed.) DHEW Publication No. ADM 77-328. Washington, D.C.: U.S. Government Printing Office, 1977.

Michael, D. *On Learning to Plan — and Planning to Learn: The Social Psychology of Changing Toward Future-Responsible Societal Learning.* San Francisco: Jossey-Bass, 1973.

Mintzberg, H. *The Nature of Managerial Work.* New York: Harper & Row, 1973.

Patton, M. *Utilization-Focused Evaluation.* Beverly Hills, Calif.: Sage, 1978.

Rossi, P. H. (Ed.). *Standards for Evaluation Practice.* New Directions for Program Evaluation, no. 15. San Francisco: Jossey-Bass, 1982.

Sherman, P. "Computerized CMHC Clinical and Management Information Systems: Saga of a 'Mini' Success." *Behavior Research Methods and Instrumentation,* 1981, *13* (4), 445-453.

Sherman, P., Burwell, B., and Olsen, G. *A Guidebook to the 1981 Operations Management System for Federally Funded Community Mental Health Centers.* DHHS Publication No. ADM 81-1136. Washington, D.C.: U.S. Government Printing Office, 1981.

Starling, G. *Managing in the Public Sector.* Homewood, Ill.: Dorsey Press, 1977.

Suchman, E. *Evaluative Research: Principles and Practice in Public Service and Social Action Programs.* New York: Russell Sage Foundation, 1967.

Wildavsky, A. "The Self-Evaluating Organization." *Public Administration Review,* 1972, *32* (5), 509–520.

Windle, C., and Neigher, W. "Ethical Problems in Program Evaluation: Advice for Trapped Evaluators." *Evaluation and Program Planning,* 1978, *1* (2), 97–108.

David L. Clifford is the director of planning, budgeting, and evaluation for the Washtenaw County Community Mental Health Center in Ann Arbor, Michigan. He has been affiliated with the center for the past seven years.

Paul Sherman is the associate director for administration for Ravenswood Hospital Community Mental Health Center. For seven years he served as the director of training and research with primary responsibility for program evaluation at Ravenswood Hospital.

The patterns of interaction between evaluators and managers are described, and strategies for improving relationships between the two groups are suggested.

Values and Methods: Evaluation and Management Perspectives

William D. Neigher
William Metlay

For many program evaluators working in the public supported human service system, advancement is achieved by becoming a manager. The career ladder to management seems natural enough. Program evaluators are typically professionals with the doctoral or master's degree, mainly psychologists and social workers. Most report to the agency director. They are perceived by clinical staff to be part of administration, concerned as they are with the costs of service, productivity, accountability, and statistics. In addition, the transition to management is often the only route to promotion and higher salary grades for evaluators.

This chapter does more than summarize the different competencies required for evaluation and management. It focuses on the results of the interaction between these two professional disciplines, and it stresses that what seems like an ideal marriage between two compatible fields is marked by tension, misunderstanding, and conflicting values and intentions: "Social scientists see themselves as rational, objective, open to new ideas, and committed to

The authors are indebted to Henry E. Bender for his contribution to the management section of this chapter.

A. J. Love (Ed.). *Developing Effective Internal Evaluation.* New Directions for Program Evaluation, no. 20. San Francisco: Jossey-Bass, December 1983.

truth and standards of evidence; they see decision makers as partisan, action-oriented, indifferent to evidence, irresponsible in their pursuit of quick fixes, and reluctant to consider new ideas. . . On the other side of the fence, decision makers see themselves as pragmatic, action-oriented, responsible, and informed in the ways of the world; they see social scientists as naive, jargon-ridden, oriented to esoteric academic concerns rather than to accomplishment, and irresponsible in their neglect of practical realities" (Weiss and Weiss, 1981, pp. 837–838).

The pitfalls that concern the evaluator who becomes a manager start with the age-old question of who should manage—the technician who knows every aspect of how the product is built or the professional manager who has been trained to get the job—*any* job—done right, on time, and on budget. When the job is delivery of human services during a period of major federal cutbacks, the challenge for those who administer the human services programs becomes substantially more difficult. What kind of manager is best suited to the task? What set of skills and what background are necessary? To suggest the answers, we start with an analogy to another kind of organization, where those in the ranks who have a desire to lead are given the opportunity to do so.

Consider the symphony orchestra. Orchestras and complex organizations have more in common than first meets the eye. They are both psychological groups. Deutsch (1949) points out that a psychological group exists to the extent that the individuals who compose it perceive themselves as pursuing promotively interdependent goals. Members of the psychological group engage in frequent interaction, define themselves as members, are defined by others as belonging to the group, share norms of common interest, participate in a system of interlocking roles, identify with one another as a result of the same ideals, find the group rewarding, have a collective perception of their unity, and act in a unitary manner toward their environment. Both the symphony orchestra and the complex organization share these elements.

Furthermore, the personnel of both the orchestra and the complex industrial or service organization are subject to similar chains of authority. Consider the relationships between conductor and chief executive officer, between principal chair players and top management, between section leaders and middle managers, and between symphony players and program staff. At first glance, especially in the examples from the symphony orchestra, it seems rather obvious how individuals progress vertically. To go from player to section leader to principal chair to conductor requires individual talent and assertiveness. However, exceptionally talented musicians, even exceptionally talented soloists, are not always good conductors. In addition to musical ability, a good conductor must have the skills necessary to mold individual talent into a cohesive musical whole.

The dynamics of our analogy are quite similar to instances in which human service program evaluators move into management: Their abilities, talents, and individuality and their backgrounds, values, and orientations often conflict with their new role responsibilities. To examine this issue directly, the next sections of this chapter describe the roles and responsibilities of program evaluation and program management.

The Ontogeny of Evaluators

Since others have already delineated the professional development of program evaluation (Attkisson and others, 1978; Schulberg and Parloff, 1979; Windle, 1979), we will examine the special characteristics of the profession and the forces that, so to speak, shape the evaluator, so that the conflict that invariably develops between evaluators and managers can be understood.

Clarification of these forces starts with an examination of the skills that program evaluators perceive as being critical to their profession. One survey of the judged importance of various job-related skills was conducted by Metlay and Bloom (1979). Members of the Eastern Evaluation Research Society were asked to rank order six job-related skills: management skills, writing skills, verbal communication skills, political skills, analytical skills, and substantive program knowledge. Analytical skills were rated higher in importance than other skills. Ratings favoring analytical skills and methodological sophistication were constant across different job responsibilities, such as evaluators conducting evaluation, managers of evaluation programs, and directors of organizations. Years of evaluation experience followed the same pattern as job responsibility. Furthermore, among the other five job-related functions, not only were there no significant differences, but the variability in the ratings was also quite large. Although Metlay and Bloom suggest that the variability in response may reflect the heterogeneous training of those engaged in evaluation, it is significant that, despite the unique backgrounds of evaluators—in psychology, business, urban planning, or criminal justice—they were uniform in their view that methodological skills were of primary importance.

The focus on empirical, data-based methodology among evaluators to the chagrin of decision makers is legendary. Recent studies indicate, however, that social scientists and decision makers may have more in common than each attributes to the other. Weiss and Weiss (1981) found that both social scientists and decision makers evaluated the links between research and action in much the same way. When they disagreed, social scientists overestimated the role of political considerations and underestimated the role of scientific merit in decision makers' judgments about what makes research useful.

This dialectic of consistency on the one hand and of variablity on the other is characteristic not only of our attitudes toward the importance of job-related skills but of the profession's commitment toward models of academic training. With regard to variability, the evaluation community has witnessed debates on whether training should occur within a core discipline like psychology, education, or sociology; within interdisciplinary programs; or even via a centralized institute within the university that coordinated training inputs from different speciality disciplines (Schulberg and Parloff, 1979; Metlay, 1980b). Such debates, however, have not altered the commitment of academicians to training in data analysis and research design as the focus of a graduate evaluation program, wherever it is housed. In fact, examination of a series of papers and symposia (McCullough, 1975; Metlay, 1977; 1980a; Metlay and Nevid, 1981; Nevid and Metlay, 1982) that specified the details of evaluation training reveals three major themes.

First, because most evaluation projects are correlations, survey, or quasi-experimental in design, trainees should be skilled in psychometric techniques, multivariate analysis, computer science, and analytic methods that relate to issues of construct, external, and — especially — internal validity. Second, in order to provide trainees with firsthand experience in the design and implementation of field studies and in the sociopolitical realities of applied evaluative research, an internship or practicum component should be a feature of all training models. Third, it is important to stress not what the training for evaluators is but what it is not. Little or no formal training is provided in management, accounting, micro- and macroeconomics, operations research, applied social science, or the organizational structure or behavior of public or private institutions.

Thus far, we have identified two major forces that shape the values and orientations of those who engage in evaluation activities: first, the consensus that views data analysis as central to the profession; second, the social science tradition of training in research design, scientific rigor, and technical expertise. The third and final force relates to the different and often conflicting philosophies under which program evaluators carry out their mandate.

Evaluation in Practice

Evaluators, particularly those who work in government-supported social action programs, are often confronted by vague program goals, strong promises of success, and little evidence of demonstrable outcome. The programs are often aimed at highly visible or politically relevant target groups, in need or at high risk, who display a range of social, personal, and economic ills (Neigher and Schulberg, 1982). Not only are the programs themselves only partial solutions to complex problems, but often they are implemented inconsistently. Similarly, the evaluation authority contained in many federal social welfare programs describes the evaluation mission in only the broadest terms (U.S. General Accounting Office, 1982). It is in this legislative and organizational context that local program evaluators must cope with often concurrent and inconsistent rules and obligations.

For example, Windle and Neigher (1978) have identified four purposes for program evaluation: amelioration, accountability, advocacy, and evaluation research. The first purpose, amelioration, views program evaluation as generating better information for decision makers. In this view, the information produced increases the ability of staff to comprehend what they are doing, make more effective use of resources, and feel more comfortable in their decision making. The second purpose is accountability. Assuming that a program should be evaluated by the public or by those who fund and support it, this model focuses on public data disclosure and citizen participation in the evaluation process. The third purpose, advocacy, considers program evaluation as a strategy used by organizations to advance their self-interest when vying for resources. In contrast to traditional conceptions of accountability, this evaluation purpose adapts to adversarial proceedings and assumes that selective information sharing is one of the mechanisms legitimately utilized in the competition for funding. The fourth purpose is evaluation research, which applies scientific methods to establish causal linkages between interventions and outcomes.

The ability of evaluation to be timely and decision-relevant and to achieve maximum utilization is diminished when conflicting or incompatible evaluation models are applied concurrently. For example, when in-house program evaluators working at the agency level conduct program evaluation for the purpose of public accountability as required by external accrediting or monitoring agencies, the data that they generate can be at odds with the best interests of the agency itself.

Managers and Managing

This chapter is being read by individuals steeped in the discipline of program evaluation. For this reason, the emphasis will be placed on the specifics of management. A review of the management literature leads us to conclude that the term *management* has many different meanings, depending on the background and experience of the individual. Nevertheless, most people would say that management relates to getting the job done and that getting the job done usually involves other people. Peter Drucker has stated that management is a function, a discipline, a task to be done; and managers practice this discipline, carry out the functions, and discharge these tasks. We believe that these definitions are valid, but we would like to expand them by stating that management has an integrative nature. Although it consists of various components, management is an iterative process, wherein changes in one component affect the overall management equation and require changes in other components in order to keep the organization in a state of equilibrium.

An understanding of the disharmonies that exist between evaluators in the public and private sectors and the respective managers requires an understanding of the work environments in which evaluation is performed. We can portray the work environment as consisting of three components: organizational domains, organizational structure, and the management function.

The organization domains consist of the various subspecialties employed in the functioning of the organization; for example, human resources, purchasing, general and administrative services, insurance and employee benefits, and so forth. These subspecialities work together to accomplish the organization's objectives and goals. The symphony orchestra is similar to the complex organization in this respect, since it can be analyzed in terms either of families of musical instruments, such as strings, woodwind, and percussion, or in terms of specific instruments and sections, such as first or second violins or violas. These organizational domains, which include program evaluation, are comparable to the instruments that comprise the orchestra.

The second dimension is the organizational structure within which the work is to be performed. A variety of organizational structures belongs in this dimension, including matrix organizational structures and traditional structures. Those in the public arena might say that all organizational structures are the same and that these structures have been constant for extended periods of time. We contend that the structures existing in both the private and the public sectors have changed markedly over the past few decades and that these changes are predicated on changes in organizational goals and strategies. For example, public sector individuals now need to be rapidly responsive to changes in the environment. In the mental health sector, management of a halfway house requires those who are responsible for its management to be able to react rapidly and to deal in a meaningful way with various community structures. Hence, authority must be disseminated to lower levels in the organizational structure so that effective activities can be implemented in a meaningful time frame.

The third major component that influences the work environment is the management function performed at all levels of management. The basic management functions fall into six major categories (White and Broskowski, 1980; Mintzberg, 1979): Planning is a continuous process concerned with the setting of objectives and the allocation of resources needed to perform the designated activities; planning can be operational, tactical, or strategic. Organizing consists of the design and coordination of activities for the organization as well as the delegation of tasks to organizational divisions, units, or specific personnel; organizations must balance differentiation and integration. Controlling assumes that organizational goals, deadlines, and quality standards are met. Directing and leading includes the activities oriented toward leading and motivating the work force to maintain an environment of cooperation and to ensure proper implementation of the organization's policies and procedures. Evaluating makes judgments about program effort, effectiveness, efficiency, and adequacy, and these judgments are used for making data-based decisions about the organization and its future direction. Finally, financing is the process by which operating expenditures and revenues are obtained, expended, and documented.

These job functions vary according to the environment in which the work is performed, including the nature as well as the structure of the organization. Also, the percent of time that various individuals spend in any one function depends on their position within the organization as well as on the organization's objectives and goals. The manager's job often also includes communicating — the verbal and written presentation of ideas and specific activities to subordinates, peers, and superiors; staffing — the selection of department or organizational staff and the development of their skills; decision making — the conceptualization, analysis, and logical and creative use of concepts; and negotiating — the activities of conferring with others to settle matters and resolve differences.

While individual organizational tasks require a different mix of the management activities just described, they remain fundamental components of management and required skills and competencies for managers. Inherent in these skills are values that emphasize task orientation, positive outcomes, cost-effectiveness, competition, and timing. The bottom line, profit, is clearly a performance indicator for most managers who work in an industry whose goods are the tangible results of manufacture. In a services environment, customer satisfaction, market stance, market share, and profitability are the fundamental measures by which managers are judged.

Managers of publicly supported human services are often perceived as different from their counterparts in private industry. In fact, the managers of publicly supported human services organizations more often than not are former service providers who rose from the ranks to manage. Regardless of their background, however, they have learned that good management is essential to moving an organization, even when government subsidies keep the books in the black. Since as much as 80 percent of operating costs goes for personnel, human service management requires a special sensitivity to the needs of its highly skilled and pedigreed labor.

Partly because their training was not in the technical field of management science and partly because both the human service organizations and their top administration were usually of modest size, management relied heavily on other staff for help: accountants, personnel directors, and in some cases program evaluators (U.S. General Accounting Office, 1982). Ironically, in the 1970s many evaluators shared backgrounds with the managers who looked to them for support: They came into evaluation mostly from formal training in research, they had little experience in applied settings, they got much of their evaluation experience on the job, and they brought with them values and expectations that were often inconsistent with or antagonistic to effective management. In order to put these conflicting orientations into perspective, the final section of this chapter examines the competencies and values held by evaluators and program managers.

Evaluators and Managers: Competencies and Values

The two preceding sections described the management and evaluation process and the purposes and intentions served by each. This section juxtaposes the skills and competencies required for successful program evaluation and management and explores the values and orientation of the individuals who fill these roles. Table 1 defines nine competencies, which represent essential skills. The competencies are described in terminology consistent with the two disciplines. We have selected four examples to illustrate the inconsistent and even oppositional posture of similar competencies practiced for different purposes.

Central to program evaluation is a set of methodological techniques based on or derived from principles of scientific research design (Bennett and Lumsdaine, 1975; Cook and Campbell, 1979). The common purpose of these techniques is to isolate dependent variables and manipulate independent variables and to maximize the influence of independent variables while minimizing the sources of variance (Neigher and Schulberg, 1982). In other words, their purpose is to assure the reliability of implementation and to infer the validity and generalizability of a set of current findings about the particular program to a set of future outcomes from other programs. The purpose is to produce information, regardless of outcome, in a manner which assures that other investigators can replicate the results.

Formal research has rigid standards based on scientifically acceptable rules for determining success and failure; examples include $p < .05$ and Type I and Type II errors. In evaluation, findings do not have to be statistically significant in order to be decision-relevant. The consequence of evaluation findings is a point that we will address shortly, but we must first consider an analogous skill of program managers, the demonstration project.

Table 1. Evaluation and Management Perspectives

Program Evaluation		Management	
Competency	Orientation	Competency	Orientation
Research design	Replication	Innovations trials	Utilization
Management information	Data-based conclusions	Information management	Data-based decisions
Decision making	Statistics: implications	Decision making	Information: applications
Cost accounting	Description	Finance	Prediction
Needs assessment and program development	Who needs? social justice	Planning and marketing	Who wants? economic marketplace
Written communication	Inform or document	Written communication	Direct or control
Organizational theory	Understanding systems change	Organizational management	Creating systems change; motivate, achieve
Time management	Flexibility; unconstrained	Time management	Structured; limited
Interpersonal skills	Peer relations	Personnel	Hierarchical

Before large-scale adoption of a new or expensive program takes place, managers must often demonstrate on a smaller scale that the program works. The concern is for the here and now, for data that can be used to make decisions. The manager is motivated to show that a project can be done reliably and that it can achieve practical results. In contrast, program evaluators tend to look for validity with the view that the final report is not a means to a final decision but an end in itself.

Both evaluators and managers make decisions, of course, and the nature of that decision making is the focus of the next example. In evaluation, decision making is characterized by an analysis of both process and outcome, but evaluators are most comfortable with hard data shaped by statistical analyses. Evaluators are confident when their data show statistical significance; in other words, when they are publishable. However, as Berk and Rossi (1976) point out, scientific criteria are merely political decisions by another name, selected in advance and passed on by custom. Data that are not publishable are nonetheless actionable. The problems of an invalid conclusion can be specified, and the consequences can be weighed and owned.

As Patton (1978), Weiss and Weiss (1981), and others have described, decision makers use evaluation information in ways that are not always consistent with the expectations of evaluators and managers. Judgment, instinct, experience, risk management all enter the decision-making equation along with objective data. Managers must make decisions based on incomplete data, with all the facts not in and under time pressure. The longer one waits for information to make decisions, the greater the probability that competitors will have the same data and that market edge and posture will be compromised.

While a discussion of competitive market stance may seem out of place in an chapter on evaluation of public human service programs, it is important to recognize that most programs are not totally government-funded. That is, most programs require local matching funds, and most often government funds are based on decreasing formula grants given as seed money to help programs achieve self-sufficiency. To generate fees for services or maximize third-party payment and offset diminishing government support, human service programs often have to look to those who can pay. Ironically, government funds were given to programs in the first place to serve those most in need — those who had little financial, geographic, or psychological accessibility to private-sector programs. Both evaluators and managers are concerned with the recipients of their services, although, as our next example shows, they can approach the topic from different perspectives.

For the program evaluator, needs assessment is a series of methodologies aimed at identifying those most in need of services or gaps between service utilization patterns and needs (Warheit and others, 1977; Neigher and Fishman, 1980). Needs assessment is required in twenty-eight of the largest grant-in-aid programs funded by the U.S. Department of Heath and Human Services (Kimmel, 1977). In the federally funded community mental health centers program, for example, needs assessment has been identified in both the enabling legislation and in the program guidelines (Fishman and Neigher, 1979). The requirement is motivated by social justice. By identifying populations in need of services, agencies can be responsive, modify existing programs, and provide outreach services as well.

The "who needs?" orientation of needs assessment stands in increasingly stark contrast to the "who wants?" or the "who can pay?" orientation of program management. As human service agencies attempt to replace government dollars with dollars from other sources, the economic marketplace will play a greater role in determining who makes use of agency resources. In fact, the community mental health center (CMHC) programs that were most responsive to community needs often fared the worst when federal dollars ran out (Woy and others, 1981). Our point here is not that program managers are insensitive to populations in need or at risk of needing human services; rather, the program manager must strike a balance between responding only to those in need who lack the ability to pay and responding also to those who bring revenue needed to keep the organization financially viable.

In the final example, we examine the incentive systems for program evaluators and program managers and take a look at the products associated with each discipline. We present these two dimensions in Table 2. As we described in the preceding section, the reports and publications generated by program evaluators are oriented toward establishing the validity and quality of the product. In contrast, program managers are more concerned with reliability — the consistency in which agency policies and procedures are followed and the program's ability to achieve the same results time after time. On the scale between quality and quantity, the program manager needs to keep his eye on issues of quality while generating a program in sufficient quantity (productivity per unit cost) to assure adequacy and financial stability. Efforts at quality assurance, program evaluation, and continuing professional education are costly and indirect services. These efforts generate no fees in themselves. Traditionally, they have relied on government or professional support to sustain their activities. There are, in fact, very real disincentives for the delivery of high-quality programs in the public sector. Quality does not come cheap. Such

Table 2. Output and Incentives

Program Evaluation		Management	
Issue	*Orientation*	*Issue*	*Orientation*
Products: reports, publications	Validity, quality	Products: policy, procedures	Reliability, quantity
Rewards	Discover new techniques; generate information regardless of outcome	Rewards	Demonstrate established techniques; positive outcome

efforts escalate the costs of the provision of direct services, and they can result in fewer people being seen at higher unit cost. In addition, third-party payers are becoming increasingly restrictive about the kinds of quality assurance activities that are allowable for reimbursement. Again, the search is for balance, achieving adequate levels of quality for the greatest number of possible consumers in public sector human service programs.

Last, consider the reward and incentive systems for the fields of program evaluation and program management (Table 2). Evaluators are reinforced for discovering new techniques. The desire to publish, ingrained in many graduate degree programs and a major criterion for advancement in academia, is constrained by general editorial policies that restrict the publication of replication studies or studies of negative results (Fishman and Neigher, 1982). Whether a program evaluation report is published or not, the evaluator's orientation is to generate information irrespective of the outcome. Whether the findings are good, bad, or indifferent, evaluation has provided the information, and evaluation should provide a set of recommendations for others to act on. In contrast, the professional manager or administrator is rewarded for consistently demonstrating established techniques. The rewards are for positive outcomes. The administrator who does a few things very well is more salable than the administrator who has a checkered record of innovations or experiments.

In this section, we have focused for the most part on the values and orientations that are different for program evaluation and management. There are, of course, many common elements, as the parallel competencies in Table 1 show. The differences in orientation between the two professional fields are not irreconcilable. However, they do require those in each field to develop a better and more mutual understanding of each other's needs than their current practice demonstrates. For the program evaluator who is making the transition to management, a change in more than role orientation is required. There needs to be adaptation in the way in which program evaluators think, feel, and behave toward their organizations.

The conclusions that follow offer a series of recommendations to enable both program evaluators and managers to communicate more effectively with each other and to use each other's resources more effectively.

Conclusions

Many evaluation graduate studies programs require students to have practical experience in the form of an internship or field placement in an applied setting. However, this experience almost always takes place within an evaluation setting. We suggest that the field evaluation experience should be shared fifty-fifty with a supervised management experience that focuses on how decision makers use evaluation efforts in practical application. As assistant to an administrator, evaluator interns should have to recommend alter-

58

native courses of action to deal with organizational decisions and even take
responsibility for making decisions on their own where practicable.

Evaluators should also have to make action recommendations to all
levels of organization, from line staff to middle management, from top administration to governing board members. Under the usual organizational
climate, evaluation students should experience what managers experience in
routine day-to-day decision making: too little time, political demands, lack of
sufficient or reliable data, and ambiguous or severe consequences for both
action and inaction. This type of practical experience should serve evaluators
in two beneficial ways. First, it should give the beginning evaluator more than
empathy for administration; he or she can experience the real-world pressures
that decision makers face as they request, study, digest, and "utilize" evaluation results. Those who go on to direct evaluation departments should be more
responsive to the constraints under which administrators work. Second, the
field placement or internship should provide evaluators with some practical
experience and supervision in program management that they can draw on if
they move into administration at some point in their careers.

Finally, there is growing awareness that, while evaluators and program
managers often share different value systems and orientations, there is a common ground on which better working relationships can be built. In this
chapter, we have tried to describe this common ground: in the professional
training of evaluators, in the focus on specific areas of value and organizational conflict, and in recommendations that values and motives should be
clarified up front by those who manage and by those who evaluate.

References

Attkisson, C. C., Hargreaves, W. A., Horowitz, M. J., and Sorensen, J. E. (Eds.).
Evaluation of Human Service Programs. New York: Academic Press, 1978.
Bennett, C. A., and Lumsdaine, A. A. (Eds.). *Evaluation and Experiment: Some Critical
Issues in Assessing Social Programs.* New York: Academic Press, 1975.
Berk, R. A., and Rossi, P. H. "Doing Good or Worse: Evaluation Research Politically
Reexamined." *Social Problems,* 1976, 337–349.
Cook, T. D., and Campbell, D. T. *Quasi-Experimentation: Design and Analysis Issues for
Field Settings.* Chicago: Rand McNally, 1979.
Deutsch, M. "A Theory of Cooperation and Competition." *Human Relations,* 1949, *2,*
129–151.
Drucker, P. *Management: Tasks, Responsibilities, Practices.* New York: Harper and Row,
1974.
Fishman, D. B., and Neigher, W. D. "Needs Assessment: A Conceptual and Prac-
tical Overview." In. G. Landsberg, W. Neigher, R. J. Hammer, C. Windle, and
J. R. Woy (Eds.), *Evaluation in Practice: A Sourcebook of Program Evaluation Studies from
Mental Health Care Systems in the United States.* DHEW Publication ADM 78-763.
Washington, D.C.: U.S. Government Printing Office, 1979.
Fishman, D. B., and Neigher, W. D. "American Psychology in the Eighties: Who Will
Buy?" *American Psychologist,* 1982, *37,* 533–546.
Kimmel, W. A. *Needs Assessment: A Critical Perspective.* Washington, D.C.: Office of the

Assistant Secretary for Planning and Evaluation, U.S. Department of Health, Education, and Welfare, 1977.

McCullough, P. "Training for Evaluators." In J. Zusman and C. R. Wurster (Eds.), *Program Evaluation*. Lexington, Mass.: Lexington Books, 1975.

Metlay, W. "Relevance in Psychological Research: New Training Models for Research in Real-Life Settings." Paper presented at the New York Area Social Scientists Conference, December 1977.

Metlay, W. "Evaluation Research: The Emergence of a Profession." Paper presented at the New York Area Social Psychologist Conference, May 1980a.

Metlay, W. "Who Should Train the Evaluation Researchers?" Paper presented at the American Psychological Association Conference, September 1980b.

Metlay, W., and Bloom, H. "Requisite Skills and Educational Training Models for Evaluation Research." Paper presented to the Evaluation Research Society, October 1979.

Metlay, W., and Nevid, J. S. "Integration of the Real and Academic Worlds: Internship Model." Paper presented at the American Psychological Association Conference, August 1981.

Mintzberg, H. *The Structuring of Organizations*. Englewood Cliffs, N.J.: Prentice-Hall, 1979.

Neigher, W., and Fishman, D. B. (Eds.). *Needs Assessment and Mental Health Planning: A Resource Manual*. (2nd ed.) Trenton, New Jersey: State of New Jersey, 1980.

Neigher, W. D., and Schulberg, H. C. "Evaluating the Outcomes of Human Service Programs: A Reassessment." *Evaluation Review*, 1982, *6* (6), 731–752.

Nevid, J. S., and Metlay, W. "Integrating the Real and Academic World: An Experience-Based Research Training Model." *Professional Psychology*, 1982, *13*, 594–599.

Patton, M. Q. *Utilization-Focused Evaluation*. Beverly Hills, Calif.: Sage, 1978.

Schulberg, H. C., and Parloff, R. "Academia and the Training of Human Service Delivery Program Evaluators." *American Psychologist*, 1979, *34*, 247–254.

U.S. General Accounting Office. *A Profile of Federal Program Evaluation Activities*. Washington, D.C.: Institute for Program Evaluation, U.S. General Accounting Office, 1982.

Warheit, G. J., Bell, R. A., and Schwab, J. J. *Planning for Change: Needs Assessment Approaches*. DHEW Publication ADM 77–472. Washington, D.C.: U.S. Government Printing Office, 1977.

Weiss, J. A., and Weiss, C. H. "Social Scientists and Decision Makers Look at the Usefulness of Mental Health Research." *American Psychologist*, 1981, *36*, 837–847.

White, S. L., and Broskowski, A. "Critical Management Tasks: The PERFORM Model." In S. L. White (Ed.), *Middle Management in Mental Health*. New Directions for Mental Health Services, no. 8. San Francisco: Jossey-Bass, 1980.

Windle, C. "Developmental Trends in Program Evaluation." *Evaluation and Program Planning*, 1979, *2*, 193–196.

Windle, C., and Neigher, W. "Ethical Programs in Program Evaluation: Advice for Trapped Evaluators." *Evaluation and Program Planning*, 1978, *1* (2), 97–107.

Woy, J. R., Wasserman, D. B., and Weiner-Pomerantz, R. "Community Mental Health Centers: Movement Away from the Models." *Community Mental Health Journal*, 1981, *4*, 265–276.

William D. Neigher is assistant director of Mental Health Services at St. Clare's Hospital, New Jersey, and visiting associate professor of psychology at Rutgers University Graduate School of Applied and Professional Psychology. A past president of the Eastern Evaluation Research Society, he has been a frequent consultant to the federal government and various state governments.

William Metlay is a professor and director of doctoral studies in applied research and evaluation in psychology at Hofstra University. He is a consultant to numerous organizations on management techniques, systems analysis, and program evaluation.

Effective internal evaluation requires dependable data for decision-makers. Influences on data dependability are discussed.

Influences on Internal Evaluation Data Dependability: Organizational Issues and Data Quality Control

Frederick L. Newman
Richard White
Deborah Zuskar
Eric Plaut

Internal evaluation efforts do not achieve their goals unless internal evaluators can address certain key issues. Are the data sufficiently reliable to be used in making management decisions? What will cause the reliability and validity of the data to deteriorate? How can one know when these factors have rendered the data no longer believable? What precautions can be taken to maintain or increase the dependability of data?

The theme of this and the following chapter is that it is possible to maintain clinical and fiscal data reliability at a level that is sufficiently high to render the data useful for internal evaluation provided that certain minimum precautions are taken in four areas: organizational influences, data quality control, staff concerns, and staff and patient variables. Within the organizational structure, evaluation responsibilities must be spelled out: Who is

A. J. Love (Ed.). *Developing Effective Internal Evaluation*. New Directions for
Program Evaluation, no. 20. San Francisco: Jossey-Bass, December 1983.

responsible for evaluating what, with which data, with what authority, and with what range of potential actions? The mechanics of data collection and processing need to be identified and documented so that they can be monitored and controlled. The concerns of staff about evaluative questions and the data collection procedures on them need to be identified and addressed. Finally, the staff and patient variables that can bias the data need to be identified and minimized.

In this chapter, we discuss what the literature and our own experience recommend for the first three areas. In Chapter Five, we address the fourth area, and we describe the results of studies investigating the influences of staff training and experience on data dependability.

Organizational Influences

The question of organizational influences raises two issues. The first is one of identifying the evaluation responsibilities of all staff, from line staff to supervisors, managers, and members of the governing board (Elpers and Chapman, 1978; Newman and Sorensen, in press). We define evaluation responsibilities in terms of the kinds of decisions—and the potential consequences of these decisions—that define responsibilities at each level of the organization. Stated another way, the questions that need to be addressed at each level of the organization have to be defined, and the data that need to be collected so that these questions can be answered have to be identified (Bieneman, 1971; Newman and Sorensen, in press). The second issue can be described as one of the influence of data on the organizational authoritative structure: How much clout do the evaluative data have in influencing staff roles, activity, and survival (Weiss and Bucvalus, 1980)?

The evaluative questions of managers will influence the meaning and use of the information that is gathered (Finn and Miller, 1971; Newman and Sorensen, in press). For example, questions that focus on the amount of direct service, regardless of patient problems and goals, will indicate to staff that management is interested in time productivity, not in the types of client and patients served or in the clinical outcomes. Conversely, an organization in which questions focus not on service resources consumed and revenues generated but on patient problems, goals, and outcomes will eventually find itself in financial difficulty. Our bias is that, for the most part, all management questions should relate to our generic question, which is then adapted appropriately to each of the organization's levels, from front-line service staff to executive director.

The generic question is, What is the relationship between client patient characteristics, the clinical procedure applied, and the patterns of clinical outcomes, resources consumed (costs), and revenues that result? For service staff, the question could be a treatment planning issue that translated as: What combinations of individual, group, and adjunctive therapy will be most effective for a patient with borderline personality structure, poor social supports, and a middle-to-low level of daily functioning and also stay within the cost reimbursement constraints of Blue Cross? For program managers, the evaluative question could translate as: What arrays of clinical services and clinical staff resources should be available to the service unit so that it can provide the array of services needed by borderline patients with poor social supports and moderately low levels of daily functioning, given the available sources of revenue generation?

Once the generic evaluative question has been formulated at each level, a whole set of specific evaluative questions needs to be formulated. For example, the overriding generic question just given for clinical staff implies a set of more specific questions, such as these: What types of individual and group therapy techniques can best accomplish the goals that I have formulated for a given patient within the time and reimbursement constraints imposed by the third-party payer? How will I recognize improvement or regression? What monitoring and review strategy will best coordinate the efforts of those who provide the various therapeutic interventions?

Formally, the family of evaluative questions generated at the various levels of the organization will require data to be captured under five domains: client and patient characteristics (demographic and clinical); clinical procedures and processes; clinical outcomes; costs of resources consumed in the clinical processes; and revenues (or potential revenues) from first- and third-party payers. Client and patient characteristics and clinical procedures are typically the independent or predictive variables in evaluative questions. Clinical outcomes, costs, and revenues are the dependent variables in the evaluative questions. Data elements within each of these five domains must be defined in such a way that they will be understood and useful to frontline clinical staff. Stated another way, the data elements must relate directly to the observations that clinical and clerical staff can record reliably. The data collected by clinical staff become the data used by people at all levels of the organization for internal evaluation purposes. This brings us to the issue of how data are used for organizational authority and control.

It is painfully obvious that the data collected by frontline staff and monitored and controlled by middle-level staff are used by middle- and upper-level management staff to shape the role responsibilities and authorities of people at all levels of the organization (Kadin and Green, 1971). The focus and substance of evaluative questions are necessarily shaped by the level of responsibility and authority that is granted to each level. Yet, the greatest responsibility for data collection and processing is bestowed on frontline staff, while the greatest responsibility for assuring the reliability and validity of data is given to supervisory or middle-level staff.

Decisions evaluating the impact of staff roles and activities are based in the main on data collected or generated by people who are closely involved in the day-to-day operation of clinical processes. On the one hand, frontline staff can be concerned about how the data that they collect will influence their survival and status in the organization. On the other hand, management staff will be concerned about whether they can trust the data generated by staff whose activities they are attempting to understand in terms of cost outcome and revenue results. These concerns lead us to focus next on data collection procedures and mechanics and on staff concerns.

Data Quality Control

Data collection is a trial-and-error process unless a well-documented plan can be devised, tested, and maintained by carefully devised monitoring procedures. The plan should be documented in both written and diagram form. The plan should give an overall description as well as detail the sequential steps involved in data collection and monitoring, and the diagram should identify which staff are to capture each of the data elements and in what sequence. In the absence of such a plan, the ability to monitor and control data collection becomes a trial-and-error process (Neel, 1971).

It is easy for staff to impose their own perspectives on the data collection process, thereby increasing the variability in the data captured and decreasing their reliability and validity. Thus, instructions must classify the data elements into mutually exclusive and exhaustive sets that staff responsible for capturing the data can discriminate easily. The instructions must also clearly describe the steps in which the data are to be compiled on printed forms or into the computer terminal. Explicit instructions reduce the probability of variability in the interpretations of staff who record the data.

The final draft of the diagram should include a monitoring process that identifies who is responsible for monitoring the reliability of the information collected. One important element in the development of the monitoring process is to provide it with clout. For example, while it may be cost-efficient to have a low-salaried clerical staff person monitor the process, it is also important to identify a supervisor or program director who will enforce the data collection standards. As part of this process, clear and unambiguous sanctions of disregard for data collection procedures should be identified.

Usually, several weeks (in some instances several months) of pilot testing are required before data elements can be defined adequately and before it can be demonstrated that the data collection forms are reliable instruments. Pilot testing should measure both the accuracy of the data collected and the time or effort required. How expensive is the data collection effort to those involved in collecting the data? How much does the data collection effort detract from the activities that staff value? During the pilot testing period it also is important to monitor staff concerns about the data that they are collecting. When staff see data lacking face validity and usefulness or as undesirably exposing the tough clinical decisions that they have to make, it often shows up in pilot testing. We will have more to say about this topic later in this chapter.

Training in the proper collection and use of data is another necessary aspect of the pilot testing period. What type of training and how much is necessary to initiate the data collection procedures? Often, the training can be integrated into the usual staff supervision procedures. One form of training involves an exercise in which all staff of one service unit independently review and complete data collection forms on a small set of cases presented in the form of written, audio taped, or videotaped transcripts. Clinical staff and supervisor can then review the response patterns of service unit members. Differences in responses can be due to problems created by the definitions and instructions, or they can be due to differences in the frames of references of staff. Such differences provide a basis for determining the extent to which staff are using a common frame of reference.

Finally, one person must be given the responsibility and authority to coordinate the implementation and maintenance of the data collection and processing system. This person should have an advisory committee that represents all the major service and administrative components of the organization. The role of the committee is to review and monitor the structure and process of the information system. Its goal is to recommend modifications to the system that can improve its efficiency in addressing the evaluating questions required to operate the various service units of the organization.

Staff Concerns

Staff concerns about how data will be used in the organization's decision making greatly influence the reliability and timeliness of data collection efforts. Consider some examples of staff concerns: Who uses the data that describe my activities, and how are these data used? Will the data help me to do what I get paid to do? Will the data describe the impact of what I or my staff do with client and patients accurately? These are typical concerns of staff at any level of the organization's hierarchy, from line staff to executive director (Newman and Sorensen, in press; Wilkinson, 1974).

Forquer and Anderson (1982; Anderson and Forquer, 1982) describe how such concerns can influence the implementation of data collection procedures in treatment planning and quality assurance review. Their study indicates that an ordinal array of staff concerns surface when treatment planning and quality assurance procedures were being developed and implemented. Since data collected for treatment planning and quality assurance review are often the same as those used for internal evaluation, their results are germane to the present discussion. Forquer and Anderson found that implementation and proper use of the procedures correlated directly with one of three sequential levels of staff concerns: Self-concerns, task concerns, and impact concerns. Self-concerns are concerns about how the procedures and the resulting data can influence the staff's survival in the organization and the role and status that they have in the organization. Task concerns are concerns about how to perform the tasks required by the procedures at a level of competency that satisfies both the supervisor and the staff person. Impact concerns are directed toward outcomes and dissemination and refinement of the procedures being implemented. Impact concerns, in turn, are subdivided into two subsets, differentiated by whether implementation of the procedures results in good outcomes or in bad outcomes for the organization. If the results of the procedures are perceived by management as being helpful to the organization, the concerns of staff take the form, Do I get credit and reward for my role in implementing the procedures that help us to look good? However, if the results of implementing the procedures are perceived by management as being harmful to the organization, then the concerns of staff take the form, To what extent do I bear responsibility and to what extent do others share responsibility if implementing the procedures has some negative consequences?

Forquer and Anderson found that they could trace the level to which the quality assurance and data collection procedures were implemented to the level of concern voiced by both frontline and management staff. If self-concerns were prominent, then the implementation process usually bogged down in discussion about why the procedures should be implemented and whether they had value. Here are some examples of self-concerns: "The procedures take too much time and require too much paperwork." "The procedures encourage staff to play paper and number games rather than to do their clinical work." "Good paperwork and good numbers will be rewarded rather than good clinical work." If the procedures were forced into practice, the error rate was high, and the level of compliance was low.

If task concerns predominated, staff were usually attempting to delay formal implementation of the procedures. Here are some examples of task concerns: "The forms are difficult to fill out." "I do not understand how to use the quality assurance appeals procedure." "The clinical rating scale does not relate to adolescents." When task concerns were prevalent, technical staff needed to spend more time either in training staff or in refining procedures so that the procedures and the resulting data were either more accurate or easier for staff to use.

Impact concerns usually predominated during the early stages of the implementation and refinement process but only after task concerns had been addressed for the bulk of the staff. It appeared as if staff wanted to be assured of two things. First, if staff were doing the tasks appropriately, their concerns appeared to focus on whether the problems that surfaced as a result of the quality assurance and data collection procedures were going to be dealt with in a timely fashion that did not harm those who uncovered the problem. Second, they wanted to know whether the organization had the means of disseminating the results so that the results would have the greatest positive impact on both the client and patients and on the funding agenices. Here are some examples of impact concerns: "These quality assurance data collection procedures need to be coordinated with the procedures used by other agencies that service the same client patients." "The results suggest some changes that should be carefully studied in a clinical care evaluation research study." "The results indicate that staff training and closer supervision are needed for certain therapeutic techniques." "The results might force us to reevaluate our goals for some client patient groups."

When impact concerns are prevalent, a review of what William Underhill (1980) calls the organizational response is warranted. When data indicate that the organization needs to respond, its ability to respond is a function of two characteristics. The first involves the existence of procedures enabling the organization to respond. The second involves the organizational climate. Let us examine these two characteristics in turn.

The first issue is whether the organization has procedures that permit it to make an appropriate response. For example, if the data indicate that staff training is required, are there procedures to assure that the appropriate training program is designed and provided? If clinical procedures employed with a particular client patient population appear to require change, is there a way to conduct a clinical care evaluation study that could estimate whether a new procedure would be effective?

The second factor that influences the organization's ability to respond to data is what Underhill describes as the organizational climate. How does the organization encourage or discourage the use of data collected from procedures such as those of the quality assurance program implemented by Anderson and Forquer? For example, when the only relevant data for survival answer the question, "What does the boss think to be correct?" then the data collected from quality assurance procedures will have little relevance to bringing about change. However, if the organizational climate created by administrative and clinical leadership encourages interaction of data with management experience, then impact concerns will fuel efforts to improve current procedures and provide more cost-effective services.

The staff concerns just described are generic issues in the development of any internal evaluation process, whether the process evaluates human services or manufacturing. However, some issues are of particular concern to the staff working in mental health service settings. Three concerns require special attention. The first is the issue of confidentiality. If a therapist does not share information provided directly by the client patient with anyone else, then confidentiality is assured. However, when clinical information is placed in a manual or computerized information system, the therapist can only trust the reliability of the system. The larger the system and the more data computerized, the greater the tendency for clinical staff to distrust the system's ability to assure confidentiality. If computerized support of quality assurance and internal evaluation procedures is more efficient, then it becomes more difficult to guarantee that clinical data will be used only for intended purposes and that access to them can be controlled.

The second issue is the therapist's responsibility to the client patient.

68

The therapist is not just the agent of the employer. He or she is also, both ethically and legally, the agent of the client patient, and he or she is responsible to and for the client patient. In an ideal world, the two sets of responsibilities would not conflict. In the real world they often do, or at least the potential for conflict is perceived to exist. As a result, therapists and other service staff may be guarded about some aspects of their work with client patients. The issue is whether, and to what extent, quality assurance and internal evaluation procedures infringe on this responsibility.

The third issue is one of how the results of internal evaluation are used to shape the course of clinical work. Evaluation data can be seen as reflecting on the highly personal nature of the psychotherapeutic work. Because the personal alliance between therapist and client patient is central to the therapeutic process, therapists can feel that it is not only their work that is being evaluated but also their personal identity and value as human beings. As a result, good clinical supervision can take place only in an atmosphere of personal trust. Because line supervisors not only have authority over supervisees but different areas of responsibility as well, there are limits to the trust possible in the relationship. The highly personal nature of much psychotherapeutic work further narrows those limits. Thus, there is likely to be conflict between the administrative supervisory function, which makes use of the results of internal evaluation, and the clinical supervisory function, which relies on personal trust. The resolution of such conflict is not clear. It is clear, however, that the distance between the experience and views of the supervisor and of the supervisee can enhance or impede efforts to deal with such conflict. The greater the differences between supervisor and supervisees, the greater the potential for conflict, which in turn may influence the reliability and validity of data collected for internal evaluation procedures.

The bottom line in dealing with staff concerns is that there must be procedures for bringing staff concerns about the data collection and internal evaluation procedures to the surface. Moreover, there must be procedures for addressing these concerns. The procedures must be appropriate to the particular level of concern that predominates—self, task, or impact. Moreover, the organizational climate in which the concerns are expressed must also be considered. For more discussion of these issues, the reader can consult the references cited in Anderson and Forquer (1982) and Forquer and Anderson (1982), particularly Jesse (1981), Argyris (1975), and Havelock and Havelock (1973).

References

Anderson, T., and Forquer, S. L. "Use of a Concerns-Based Technical Assistance Model for Quality Assurance Implementation." *QRB/Quality Review Bulletin,* 1982, *8* (12), 4–11.

Argyris, C. *Intervention Theory and Method: A Behavioral Science View.* Reading, Mass.: Addison-Wesley, 1975.

Bieneman, J. N. "The Computer in a Management Information System." In W. C. House (Ed.), *The Impact of Information Technology on Management Operation.* New York: Auerbach, 1971.

Elpers, J. R., and Chapman, R. L. "Basis of the Information System Design and Implementation Process." In C. Attkisson, W. Hargreaves, M. Horowitz, and J. Sorensen (Eds.), *Evaluation of Human Service Programs.* New York: Academic Press, 1978.

Finn, K. R., and Miller, H. B. "Is Your MIS Fit for Human Consumption?" *Industrial Engineering,* 1971, *3* (11), 18-20.

Forquer, S. L., and Anderson, T. "A Concerns-Based Approach to the Implementation of Quality Assurance Systems." *QRB/Quality Review Bulletin,* 1982, *8* (4), 14-19.

Havelock, R. G., and Havelock, M. C. *Training for Change Agents.* Ann Arbor: Institute for Social Research, University of Michigan, 1973.

Jesse, W. F. "Approaches to Improving the Quality of Health Care: Organizational Changes." *QRB/Quality Review Bulletin,* 1981, *7* (7), 13-18.

Kadin, M. B., and Green, R. "Computerization in the Medium-Sized CPA Firm." *Journal of Accountancy,* 1971, *17,* 44-49.

Neel, C. W. "Computer Conduct in Mechanical Systems." *Journal of Systems Management,* 1971, *22* (12), 35-38.

Newman, F. L., and Sorensen, J. E. *Integrated Clinical and Fiscal Management in Mental Health.* Norwood, N.J.: Ablex, in press.

Underhill, W. "Organizational Climate and the Management Use of Data." Paper presented at the annual meeting of Technical Assistance Providers, New Orleans, 1980.

Weiss, C. H., and Bucvalus, M. J. "Truth Tests and Utility Tests: Decision Makers' Frames of Reference for Social Science Research." *American Sociological Review,* 1980, *45,* 302-313.

Wilkinson, J. W. "Guidelines for Designing Systems." *Journal of Systems Management,* 1974, *25* (12), 36-40.

Frederick L. Newman is coordinator of research, evaluation, and information systems at Northwestern Institute of Psychiatry. He has conducted program evaluations for over fifteen years in a wide variety of settings.

Richard White is coordinator of information systems at Northwestern Institute of Psychiatry.

Deborah Zuskar is research project director for alternative care at Northwestern Institute of Psychiatry.

Eric Plaut is vice-chairman of the Department of Psychiatry and Behavioral Science at Northwestern Institute of Psychiatry.

Methods for identifying and reducing clinician factors that bias internal evaluation data are presented.

Influences on Internal Evaluation Data Dependability: Clinicians as a Source of Variance

Frederick L. Newman
Mary Ann Heverly
Mitchell Rosen
S. Mark Kopta
Rebecca Bedell

The extent to which internal evaluation data are useful in clinical and program management decision making is related to the program's capability to evaluate two issues. The first issue is the ability to identify whether, and how, critical factors bias the data. The second is the ability to make adjustments that correct for the bias. If one can detect whether particular factors are influencing the deterioration in data reliability, then one can remedy the problem in one of two ways. One remedy is to adjust the data statistically for the biasing covariates. The other remedy is to create administrative procedures that significantly inhibit the contaminating sources of variance from occurring. What is needed is a means of surfacing the factors that bias the data.

This research was supported by NIMH Grant 37038, awarded to Frederick L. Newman.

A. J. Love (Ed.). *Developing Effective Internal Evaluation.* New Directions for
Program Evaluation, no. 20. San Francisco: Jossey-Bass, December 1983.

This chapter performs two functions: First, it describes a set of procedures that can be used to surface clinician factors that interact with client and patient variables to increase the variance in cost outcome data. Second, it reports the results of a relatively large study that used the procedures described. The results of this study provide guidelines to sources of variance that may influence the dependability of cost outcome data in other internal evaluation studies. The study reported here sought to identify factors influencing clinical data dependability to five community mental health centers (CMHCs) and two state hospital programs in New Jersey. In discussing the results, we introduce data from other research that applies to the issue of clinical data dependability.

The purpose of the New Jersey Clinical Data Dependability Study was to determine whether factors found to be significant sources of variance in actual client and patient cost outcome results interacted with four clinician-related variables in clinician rating of client and patient level of functioning and in clinicians' treatment planning recommendations. Previous research indicated that four clinician variables could be expected to influence the clinicians' level of functioning ratings and treatment planning recommendations. These four variables were experience (years of clinical work in mental health), degree (doctorate, master's, bachelor's, no bachelor's degree), theoretical orientation (psychodynamic, cognitive-behavioral, family systems, psychosocial, other), and service setting (inpatient, partial hospitalization, outpatient; CMHC, state hospital).

We anticipated that the variables of experience and degree would be directly related to increased reliability in level of functioning ratings. As a result of earlier work (Newman and Rinkus, 1978), we expected a significant interaction between the clinicians' experience and degree and their ratings of client patient level of functioning and that this interaction would correlate with differences in treatment recommendations. That is, the tendency to rate people at lower levels of functioning due to nonsymptomatic psychiatric factors such as social isolation, unstable employment, and race or enthnicity, would be inversely related to the restrictiveness and cost of the treatment recommended.

Our concern about clinicians' theoretical orientation was derived from recent research (Giladi and Newman, in press) describing how clinicians of different theoretical orientation differed in their use of clinical material in formulating clinical assessments and recommending therapeutic interventions. However, the client patients rated by clinicians in the Giladi and Newman study had a very narrow set of clinical characteristics (middle-class people with moderately severe depression). We wondered what would occur with client patients who exhibited a wider range of clinical characteristics? Would the differences due to clinician's theoretical orientation found by Giladi and Newman be constant across client patients who characteristically differed in their cost outcome results? We did not think that they would. For example, such client patient variables as level of social support and cooperativeness with treatment could have a different impact on the clinical assessment and treatment planning decision of psychodynamic, family systems, and cognitive-behavioral clinicians. A procedure should be able to detect whether such bias exists, and if it does, in what direction and to what degree. If we had such information, we could enter therapist's theoretical orientation into a regression equation as a covariate and adjust cost outcome results statistically.

Finally, we anticipated that the clinician's work setting would influence both the clinician's judgments of client patient level of functioning and the clinician's treatment recommendations. Here, we asked whether clinicians working in more restrictive treatment settings, such as inpatient or state hospital settings, would tend to recommend more restrictive treatment plans and show interactions in their level of functioning ratings as a function of the client patient's prior hospitalization history.

We encountered one major problem in designing a research strategy that addressed these evaluative questions directly. The New Jersey data collection system did not permit us to cross-reference clinicians' identification with client patient records, so another research strategy had to be developed. We decided to divide the study into two parts. In Part A, we identified the client patient variables that were related to differential cost outcome results in client patients served by five community mental health programs. In Part B, we used a specially constructed set of written case vignettes to identify whether the four clinician variables interacted significantly with the clinicians' ratings of client patient level of functioning and treatment recommendations. The specially constructed case vignettes were designed to differ along the client patient variables that Part A had shown to be related to actual client patient costs outcome results. If differences in clinician characteristics accounted for significant sources of variance in judged level of functioning and in treatment recommendations on the experimenter-controlled case vignettes, we reasoned that the same sources of variance could be influential in actual client patient cost outcome results.

For the sake of research cost containment, we elected to focus our analyses on male client patients between the ages of twenty-five and fifty, served by the five CMHCs and two state hospitals between 1979 and 1981. The issue of cost containment arose in Part B because the evaluation research questions that we addressed required us to control the information contained in the written case vignettes across six variables. To add gender as a seventh fixed effect would have doubled the number of case vignettes rated by clinicians. Thus, in the formal sense, the current study can only be generalized to the population of male client patients served in the public mental health system.

Part A: Identifying Client Patient Characteristics Related to Cost Outcome Results

The client patient variables under investigation were selected because of their significant effects on the variance in outcome and treatment cost results of 954 actual male client patients, ages twenty-five to fifty, served by the five community mental health programs. Table 1 describes the results of a stepwise multiple regression analysis performed on two sets of actual client patient dependent measures. First, we analyzed the influence of client patient predictor variables on the dependent measures of admission and discharge level of functioning. The level of functioning scale used by the New Jersey Division of Mental Health was adapted from a global scale developed by a mental health program in Montgomery County, Pennsylvania (Newman, 1980). It was predicted from other research (Newman, 1983), which showed that five variables could interact significantly with client patient functioning: home living arrangement (assumed to be level of social support), employment history (stable or unstable), prior hospitalization history (state hospital, state and community hospital over time, community hospitalization only, or no prior hospitalization), cooperativeness with treatment, and race or ethnicity (black, Hispanic, or white).

As the first two columns of Table 1 show, five independent variables produced significant interactions with admissions and discharge level of functioning. The results indicated that the relationship of levels of all five variables was differentially related to outcome as it interacted with initial level of client patient functioning. Stated another way, the outcome for a particular group of client patients was not independent of the particular combination of admissions level of functioning and level of each of the five independent variables. For four of the five independent variables, the negative characteristic was associated with lower admission levels of functioning. The reverse was true for the cooperativeness variable, where uncooperative client patients were seen, on the average, to be functioning at a slightly higher level on admission. Positive client patient characteristics were associated with significantly higher discharge level of functioning ratings. Moreover, the amount of change was significantly greater at the positive client patient characteristic level. The one exception was the cooperativeness variable, which showed no interaction effects. The issue that still needs to be addressed is whether these effects are due to the characteristics of these client patients or to a rating bias among clinicians.

We also investigated the influence of the five client patient independent variables on two dependent measures that sought to estimate the amount of therapeutic effort expended on the client patients. The first estimate was weighted length of treatment, and the second was treatment costs. For the weighted length of treatment measure, we multiplied the number of units provided of a given service by a constant representing the rank order of the level of restrictiveness of that service. Each day of inpatient service was multiplied by three, each day of partial hospitalization service was multiplied by two, and each outpatient visit was multiplied by one. By summing the products of service weighting constant times the number of units of service, we obtained a single measured value of treatment effort, weighted for the level of restrictiveness of the services provided. The higher the value of this measure, the greater the effort expended.

Table 1. Males, Ages Twenty-Five to Fifty, Served by Community Mental Health Programs

| | Dependent Variables | | | |
| | Level of Functioning | | Weighted Length | Treatment Cost |
Independent Variables	at Admission	at Discharge	of Treatment	Estimates[a]
Social Support A				
(Problem of Social Support)				
— Low	5.97	6.48	9.00[b]	500.00[b]
— High	6.66	7.16	33.19	3,200.00
	(no interaction effect)		$p < .001$[c]	$p < .001$
	(no main effect $p < .05$)			
Social Support B				
(Home Living Arrangement)				
— Foster care group home	4.91[b]	5.00[b]	271.57[b]	4,720.71[b]
— Lives alone	6.50	6.98	36.45	2,069.43
— Lives with family	6.89	7.42	21.22	1,402.56
	$p < .05$		$p < .001$	$p < .001$
	(main effect $p < .001$)			
Employment History				
— Unemployed	5.75[b]	6.09[b]	171.57	1,861.78
— Not in work force	5.58[b]	5.84[b]	206.27	2,047.88
— Employed	7.35	7.98	38.70	1,492.85
	$p < .01$		$p < .001$	not significant
Prior Hospitalization				
— State hospital	5.28[b]	5.50[b]	187.44[b]	4,047.00[b]
— State and community hospital	5.38[b]	5.70[b]	48.48	952.50
— Community hospital	6.10	6.17[b]	19.19[b]	512.96
— No history of hospitalization	7.11	7.65	65.92	1,375.84
	$p < .001$		$p < .001$	$p < .001$
Cooperativeness				
— Uncooperative	6.88	7.03[b]	21.48[b]	6.36.67[b]
— Cooperative	6.60	7.34	34.22	2,247.20
	(No main effect)		not significant	$p < .001$
	(significant interaction)			
	$p < .001$			
Race or Ethnicity				
— Black	5.84	6.12	92.37[b]	3,363.80[b]
— Hispanic	6.78	6.90	12.16	6.17.11
— White	6.84	7.40	28.13	1,497.70
	$p < .01$		$p < .001$	$p < .001$

[a] The error variance for treatment costs estimates was large (equal to or greater than the mean) and heterogeneous among groups. Error variance magnitude was related to the number of patients to have one or more weeks of inpatient care.

[b] These variables were significant contributors to a stepwise multiple regression equation ($p < .05$), with each variable encoded as a dummy variable.

[c] Level of functioning p values are for main effects and interactions unless stated otherwise.

For the treatment costs measure, we multiplied the number of units of a given service by an estimated unit of service cost. The unit of service cost values were set by a panel of consultants from the mental health centers. These values, which were representative of costs in fall 1982, were as follows: $230 per inpatient day, $30 per partial hospitalization day, $45 per individual therapy visit, $27 per family therapy session, and $22.50 per group therapy session. By multiplying the number of units of a given service provided during a client patient's clinical episode by the estimated costs of that service and then summing the costs over the services provided, we obtained an estimate of treatment costs. We assumed that this measure was correlated with the amount of mental health resources consumed.

Either one or both of the two treatment effort dependent measures were significantly related to all five client patient independent variables. These results are shown in the last two columns of Table 1. Five client patient characteristics were related to higher costs and more restrictive treatment efforts: a foster care or group home living arrangement (followed by living alone), not being in the work force, hospitalization in a state hospital, cooperativeness with therapeutic intervention, and black ethnicity. It should be noted that these five client patient characteristics were not independent of one another. While each contributed to the overall variance in the stepwise multiple regression equation ($p < .05$), only home living arrangement contributed more than 3 percent of the variance independent of the other variables, and the characteristic of black ethnicity accounted for less than 1 percent of the variance, independent of the other variables. Once again, interpreting the results of the analyses of the interaction of client patient variables with the two measures of treatment effort raises some problems. We need to address the extent to which these effects can be attributed to differences in clinician characteristics and not to differences in client patient characteristics.

Part B: Identifying the Influence of Clinician Variables on Level of Functioning Ratings and Treatment Recommendations

A sample of 174 clinicians from the five community mental health centers and the two state hospitals was asked to rate each of the eighteen written case vignettes along two principal dimensions: level of functioning and amount of treatment in the "most important treatment modality" needed to get a client patient through his clinical episode successfully. Clinicians were also asked to identify the goodness of the information needed to make these two judgments on a five-point scale. One represented "not at all useful," and five represented "perfectly adequate." The objective was to provide enough information in the vignette to allow the clinicians to make a reliable judgment but not enough information to eliminate sources of variance due to judgment bias. That is, we aimed to obtain average goodness of information ratings clustered around the midpoint of three on the five-point scale.

All eighteen case vignettes were constructed to represent close to the same level of moderately severe depression (Heverly and others, in press), but they were also constructed to vary along six client patient characteristics: level of social support (low or high), level of employment stability (unstable or stable), level of cooperativeness with treatment (uncooperative or cooperative), race or ethnicity (black, Hispanic, or white), prior hospitalization (state hospital, community center, or no hospitalization), and depression symptom type (external or internal symptoms of aggression). A separate group of judges (graduate students at two local colleges and clinical colleagues) rated the sentences in the written case vignettes that differentiated each of the six client patient variables to determine that the sentences did in fact differentiate the cases along these six dimensions. The procedures used are described by Heverly and others, (in press).

Results. The overall objective of Part B was to investigate the interaction of clinician characteristics with client patient characteristics in the rating of client patient level of functioning and in treatment recommendations. This investigation required us to address two questions. First, was there evidence that the clinicians were reliable in their ratings of client patient level of functioning? Second, did the main effects of the client patient variables controlled in the vignettes produce the same overall effects that we found in the actual client patient results?

Level of Functioning Scale Reliability. Scale reliability was estimated by analyzing the interrater correlations across cases. Interrater reliabilities were analyzed within specific subgroups to identify the subgroups that contributed to decreases in scale reliability. The overall reliability was high for all clinician subgroups, with one exception: Clinical staff at the state hospitals who did not regularly use the level of functioning scale in their clinical record keeping on client patients had a reliability coefficient of 0.57, while the reliability for state hospital staff who regularly used the scale was 0.71. The community program staff showed an overall reliability coefficient of 0.73. The lowest level of reliability among community program staff was exhibited by those with a bachelor's degree or less ($r = 0.68$). Those with the master's and doctorate degrees had coefficients of 0.73 and 0.81, respectively. These results indicate that training and supervision of less experienced staff in the use of the level of functioning scale should result in higher levels of scale reliability.

Main Effects of Differences Due to Case Vignette Characteristics. We were very successful in obtaining goodness of information ratings on the eighteen case vignettes within the middle range of 2.5 and 3.5 with 95 percent confidence. Results of the analysis on controlling for client patient main effects indicated that we were less successful there. Three of the six main effects were significant across all three dependent measures: cooperativeness with therapeutic intervention (cooperative or uncooperative), social support (high level of family and friend support or apparent social isolation), and employment stability

Table 2. Main Effects of Vignette Variables

Independent Variables	Dependent Variables Adjusted Means (Standard Error of the Means)		
	Level of Functioning	Weighted Length of Treatment	Recommended Treatment Costs
Symptom Type			
Internal	5.90 (.26)	80.57 (9.18)	3188 (688)
External	5.94 (.26)	70.50 (9.18)	2683 (688)
	$p = .916$	$p = .484$	$p = .638$
Cooperativeness			
Cooperative	6.50 (.30)	54.22 (10.73)	1211 (804)
Uncooperative	5.34 (.30)	96.85 (10.73)	4649 (804)
	$p = .043$	$p = .04$	$p = .030$
Social Support			
High	6.59 (.27)	51.66 (9.73)	1585 (729)
Low	5.25 (.27)	99.41 (9.73)	4286 (729)
	$p = .012$	$p = .012$	$p = .042$
Race or Ethnicity			
Black	5.51 (.35)	84.91 (12.43)	4046 (931)
Hispanic	6.06 (.32)	69.01 (11.61)	2114 (870)
White	6.19 (.30)	72.68 (10.79)	2647 (808)
	$p = .392$	$p = .692$	$p = .412$
Prior Hospitalization			
None	6.11 (.31)	68.47 (11.28)	2961 (845)
State Hospital	6.09 (.31)	69.11 (11.28)	2312 (845)
Community Hospital	5.56 (.31)	89.03 (10.48)	3534 (785)
	$p = .359$	$p = .334$	$p = .587$
Employment Stability			
Stable	7.14 (.28)	25.01 (10.10)	476 (757)
Unstable	4.69 (.28)	126.06 (10.10)	5394 (757)
	$p = .0004$	$p = .0001$	$p = .003$

(history of steady employment over several years or history of unsteady employment over the past two or more years).

The results of the analyses on main effects were both enlightening and discouraging. We had anticipated that the five client patient characteristics in evidence in the actual client patient data would have significant main effects. We had also predicted that the manifest symptoms of depression (aggression toward self or others) would be significant. The two characteristics of race or ethnicity and prior hospitalization did not show a significant main effect, nor did symptom type. These results are, to some extent, encouraging. As we already knew from the analysis of actual client patient data, the two variables of prior hospitalization and race or ethnicity covaried with (were not independent of) the other variables in their relationship to cost outcome results of actual client patients. In the study described here, these two variables were crossed more evenly with the three significant variables of social support, employment history, and cooperativeness with therapy. It is possible that the strength of these three variables in the present design attenuated the influence of the other variables. It is also possible that unemployed, socially isolated black males are overrepresented in public mental health programs relative to the general population. Thus, these two variables might only surface as significant effects when confounded with measures of social isolation and employment status.

The other nonsignificant main effect was type of depression symptom (aggression toward self or aggression toward others). We were not disappointed in the lack of significance here. This result simply confirms that level of functioning ratings and gross treatment planning recommendations, as measured in the present study (that is, by a global level of functioning scale and number of units of service) depend less heavily on the symptoms of depression that gave rise to the need for treatment than they do on other factors.

Overall, the results indicate that the level of functioning data were sufficiently reliable to permit further analyses of the interactions of clinician characteristics with the three client patient characteristics that were shown to have significant main effects.

Interaction Effects. This section focuses on the relationships between three characteristics of therapists — experience, training, and work setting — and the three client patient variables that had significant main effects. Six therapist variables resulted in statistically significant interactions of the therapist variable with as many as three client patient characteristics — employment stability, cooperativeness, and social support — on one or more of three dependent measures — level of functioning, length of treatment, and cost of treatment. The five therapist variables and the number of therapists representing each characteristic areas follows:

Theoretical Orientation:

> Psychodynamic $(n = 44)$
> Psychosocial $(n = 27)$
> Family Systems $(n = 30)$
> Eclectic $(n = 30)$
> Cognitive-Behavioral $(n = 20)$
> Other $(n = 7)$

Experience (Years of Clinical Work in Mental Health)

> Less than one year $(n = 10)$
> One to three years $(n = 38)$
> Four to ten years $(n = 85)$
> Eleven or more years $(n = 37)$

Degree

No bachelor's degree (*n* = 10)
Bachelor's (*n* = 42)
Master's (*n* = 98)
Doctoral (*n* = 28)

Service Setting

Outpatient (*n* = 96)
Partial Hospitalization (*n* = 25)
Inpatient (*n* = 38)
State Hospital (*n* = 41)
Community Program (*n* = 133)

Race

Black (*n* = 25)
White (*n* = 108)

To describe the relationship between each of these clinician variables and clinician ratings of client patient level of functioning, weighted length of recommended treatment, and estimated costs of recommended treatment, we employed a descriptive statistic called Effect Size. The effect size statistic is best understood by example. Suppose that we wish to contrast the level of functioning ratings of psychodynamic clinicians with the level of functioning ratings of family systems clinicians across the levels of the client patient variable of cooperativeness. The means and standard error of the means for this interaction are as follows:

Client Patient Characteristic

| | Uncooperative | | Cooperative | |
Therapist Orientation	Mean	Standard Error	Mean	Standard Error
Psychodynamic	4.77	.27	6.63	.27
Family Systems	5.18	.28	6.26	.28

Visual inspection of the results shows that the level of functioning (LOF) ratings across the levels of client patient cooperativeness made a larger difference for psychodynamic therapists than it did for family systems therapists. An effect size of 3.24 can be associated with this statement. This means that there was a large difference between the LOF ratings for cooperative and uncooperative client patients by psychodynamic therapists and a smaller difference between the LOF ratings for cooperative and uncooperative client patients by family systems therapists. The effect size (ES) statistic computes these differences between therapist groups across a given client patient characteristic as follows: Effect Size = ES equals (cooperative rating of psychodynamic therapists minus uncooperative rating of psychodynamic therapists) minus (cooperative rating of family systems therapists minus uncooperative rating of family systems therapists) divided by pooled standard error. Using the adjusted means and the appropriate standard error, the effect size for the two groups just described is:

$$\frac{(6.63 - 4.77) - (6.26 - 5.29)}{\text{Square Root } ((.27 \times .27 + .28 \times .28)/2)} = 3.24$$

In presenting the results, we will focus on how each of the therapist variables interacted with the three client patient characteristics across the three dependent measures. Before we describe the interaction results, we feel that some technical notes on setting criteria for assuming that the results show a significant effect are in order. Where significant interactions were observed, we then tested for the simple effects by computing the effect sizes for differences between pairs of groups over levels of the client patient characteristic (a repeated measure). If an ES was greater than 2.54, then with a probability of less than .01 (usually less than .001) we are incorrectly rejecting a true null hypothesis. Thus, with an ES greater than 2.54, we can assume that the simple effects were contributing to the significance of the overall interaction. In studying the simple effects within an interaction, we made every effort to keep the sum of the probabilities of simple effect Type I errors to be less than .05 per interaction. The logic for this criterion follows Tukey's recommendation that one should establish the probability of falsely rejecting a true null hypothesis in terms of a family of hypotheses within a larger experiment (Myers, 1979). In the current study, we consider each interaction of a therapist variable with a single client patient variable to constitute one family of hypotheses. Thus, the results reported here are confined to effect size value greater than 2.54 ($p < .01$) and with p (Type I per family of hypotheses within an overall interaction) less than .05.

Therapists' Theoretical Orientation. Significant interactions on the LOF independent measure were seen with employment stability and cooperativeness with therapy. The specific effects centered on the psychodynamic therapists. They tended to rate LOF higher for positive client patient characteristics (stable employment and cooperativeness) and lower for negative client patient characteristics (unstable employment and uncooperativeness). The simple effects contrasting differences across employment stability were between the psychodynamic therapists and three other theoretical orientations: family therapists (ES = 2.78); eclectic therapists (ES = 2.86); and cognitive behavioral therapists (ES = 3.09).

The effect sizes for client patient level of cooperativeness for LOF ratings showed that the simple effect differences in ratings across the levels of cooperativeness were between the psychodynamic therapists and the family systems therapists (ES = 3.24) and between the psychodynamic therapists and the eclectic therapists (ES = 3.19).

Therapists' Theoretical Orientation:
Level of Functioning Ratings for Employment Stability

	Stable	Unstable	Standard Error	Significant Effects			
				B	C	D	E
A. Psychodynamic	7.34	4.07	.25	−	*	*	*
B. Psychosocial	7.24	4.21	.22		−	−	−
C. Family Systems	7.05	4.49	.26				−
D. Eclectic	7.01	4.44	.24				−
E. Cognitive-Behavioral	7.14	4.69	.28				

Therapists' Theoretical Orientation:
Level of Functioning Ratings for Cooperativeness

	Cooperative	Uncooperative	Standard Error	Significant Effects			
				B	C	D	E
A. Psychodynamic	6.63	4.77	.27	−	*	*	−
B. Psychosocial	6.47	5.18	.33		−	−	−
C. Family Systems	6.26	5.29	.28			−	−
D. Eclectic	6.24	5.21	.25				−
E. Cognitive-Behavioral	6.50	5.34	.30				

The family systems therapists were typically different in their response patterns from the other four theoretical orientation groups when client patients differed in terms of their employment stability. The contrasts between the family systems therapists and the other groups netted ESs of 2.61 for the psychodynamic therapists, 3.74 for the psychosocial therapists, 3.27 for the eclectic therapists, and 5.33 for the cognitive-behavioral therapists. The psychodynamic therapists were also found to have a significant difference in length of treatment adjusted means from the cognitive-behavioral therapists (ES = 3.36).

The cognitive-behavioral therapists typically proposed less treatment

Therapists' Theoretical Orientations:
Weighted Length of Treatment and Employment Stability

	Stable	Unstable	Standard Error	Significant Effects			
				B	C	D	E
A. Psychodynamic	29.01	128.17	10.54	−	*	−	*
B. Psychosocial	24.87	143.45	14.40		*	−	−
C. Family Systems	23.48	96.33	9.58			*	*
D. Eclectic	18.03	128.75	13.30				−
E. Cognitive-Behavioral	15.13	162.18	17.21				

for cooperative client patients and significantly more treatment for uncooperative client patients than the other groups of therapists. We have inserted the effect sizes into the display that follows.

Therapists' Theoretical Orientation:
Weighted Length of Treatment and Cooperativeness

	Cooperative	Uncooperative	Standard Error	Significant Effects B C D E
A. Psychodynamic	59.83	97.34	11.19	– – – 3.24
B. Psychosocial	58.86	109.47	15.29	– – 2.59
C. Family Systems	50.64	69.17	10.18	– 5.12
D. Eclectic	54.34	92.44	14.12	3.44
E. Cognitive-Behavior	41.55	135.76	18.27	

For both the employment stability and the cooperativeness client patient variables, the cognitive-behavioral therapists consistently recommended less expensive and fewer days of treatment than therapists in the other groups for the positive characteristics (stable employment history and cooperativeness with therapy). They recommended much more expensive treatment for negative client patient characteristics in the case vignettes.

These results were sufficiently striking that we investigated further. The most expensive treatments were recommended by the six therapists who espoused a behavioral orientation rather than a cognitive orientation in describing their theoretical orientation. Four of the six worked in state hospital settings. Moreover, five of the six were female. There were no other distinguishing features; that is, all academic degrees, all races, and all amounts of experience were represented.

Therapists' Theoretical Orientation:
Estimated Costs of Recommended Treatment and Employment Stability

	Stable	Unstable	Standard Error	Significant Effects B C D E
A. Psychodynamic	283	5166	746	– – – 3.51
B. Psychosocial	406	5703	807	– – 3.06
C. Family Systems	475	4091	784	– 4.63
D. Eclectic	374	5035	857	3.58
E. Cognitive-Behavioral	109	8774	1329	

Therapists' Amount of Experience. While there were several indications for a main effect of the therapist's amount of experience in the mental health profession, only one interaction was statistically significant, the interaction of

Therapists' Theoretical Orientation:
Estimated Costs of Recommended Treatment and Cooperativeness

	Cooperative	Uncooperative	Standard Error	Significant Effects B C D E			
A. Psychodynamic	1052	4369	793	–	–	–	3.80
B. Psychosocial	1330	4779	857		–	–	3.64
C. Family Systems	1368	3198	833			–	5.06
D. Eclectic	1280	4129	910				4.08
E. Cognitive-Behavioral	592	8291	1421				

employment stability with amount of experience when measured by the length of recommended treatment dependent measure. Clinical staff who had less than one year of experience were significantly different from therapists with one to three years of experience, from therapists with four to ten years of experience, and from therapists with eleven or more years of experience. The effect sizes were 2.99, 3.49, and 2.81 for these three groups, respectively.

Therapists' Amount of Professional Experience:
Length of Recommended Treatment and Employment Stability

	Stable	Unstable	Standard Error	Significant Effects B C D		
A. Less than one year	24.56	88.36	13.43	*	*	*
B. One to three years	28.37	124.20	7.01		–	–
C. Four to ten years	21.28	127.94	10.98			–
D. Eleven years or more	28.33	127.57	11.76			

Therapists' Academic Training. While no significant interactions were obtained on LOF ratings, there was a tendency for staff with no bachelor's degree ($N = 10$) to rate client patients' LOF lower. This was a confounded effect, since the majority of these staff were state hospital personnel. As a group, state hospital personnel showed a tendency to rate client patients' LOF lower than community personnel did.

The dependent measure of weighted length of treatment had one main effect with degree, and that was with the ten clinical staff who did not have the bachelor's degree. These staff tended to recommend both more treatment and, as we will later see, more expensive treatment. The interactions with client patient level of cooperativeness showed that staff who lacked the bachelor's degree had significant effect sizes when contrasted with master's degree staff (ES = 3.23) and with staff who had the doctoral degree (ES = 3.38).

It is possible that some of the variance may be accounted for by the significant interaction of degree with race of the client patient. As the display

Therapists' Academic Degree:
Weighted Length of Treatment and Cooperativeness

	Cooperative	Uncooperative	Standard Error	Significant Effects B	C	D
A. No bachelor's	68.57	152.19	17.63	–	*	*
B. Bachelor's	56.29	104.23	12.59	–	–	
C. Master's	51.29	87.67	10.79	–		
D. Ph.D., M.D.	51.38	83.77	12.17			

that follows shows, staff who did not have the bachelor's degree recommended substantially longer treatment for black client patients than they did for white client patients (ES = 2.59). Two confounding facts will temper any firm conclusions. First, only ten respondents lacked the bachelor's degree. Second, while the majority were black, the tendency ran across white or black identification.

Therapists' Academic Degree:
Weighted Length of Treatment and Race

	Black	White	Standard Error	Significant Effects B	C	D
A. No bachelor's	132.82	91.05	19.76	–	*	–
B. Bachelor's	89.54	79.14	14.08	–	–	
C. Master's, M.S.W.	71.93	71.29	11.68	–		
D. Ph.D., M.D.	74.59	63.92	11.16			

When considering the interaction of estimated costs of recommended treatment with therapists' academic degree, staff with the master's and doctoral degrees tended to recommend less costly services, with master's degree staff recommending the least costly services for client patients with unstable employment histories (ES = 3.25 and 3.77 for the contrasts with bachelor's degree and no bachelor's degree staff, respectively). For the client patient variable of cooperativeness, there was a statistically significant effect size for the contrasts of staff with no bachelor's degree with master's degree staff (ES = 4.65) and with doctoral degree staff (ES = 3.38).

Therapists' Work Setting. There were no significant differences among clinicians in different work settings as main effects or in interactions with regard to client patient level of functioning. As for the interaction between therapists' work setting and weighted length of treatment, the source of the interaction appeared to be the fact that clinicians in partial hospitalization work settings recommended less service for client patients with stable employment and relatively more service for client patients with unstable employment histories than outpatient staff did (ES = 3.45). Inpatient staff typically recom-

Therapists' Academic Degree:
Estimated Costs of Recommended Treatment and Employment Stability

	Stable	Unstable	Standard Error	B	C	D
A. No bachelor's	3123	10647	1291	–	*	–
B. Bachelor's	312	6892	1085		*	–
C. Master's	353	4156	528			–
D. Ph.D., M.D.	101	5138	883			

Therapists' Academic Degree:
Estimated Costs for Recommended Treatment and Cooperativeness

	Cooperative	Uncooperative	Standard Error	B	C	D
A. No bachelor's	3670	11000	1371	–	*	*
B. Bachelor's	1336	5868	1152		–	–
C. Master's	1025	3484	561			–
D. Ph.D., M.D.	746	4493	937			

Therapists' Work Setting:
Weighted Length of Treatment and Employment Stability

	Stable	Unstable	Standard Error	B	C
A. Outpatient	24.94	112.32	9.96	*	–
B. Partial	16.96	141.13	11.30		–
C. Inpatient	22.88	135.90	12.51		

mended more expensive services for client patients with the negative characteristics of unstable employment history and uncooperativeness with therapeutic interventions. The effect sizes for these simple effects are given in the two displays that follow.

Therapists' Work Setting:
Estimated Costs of Recommended Treatment and Employment Stability

	Stable	Unstable	Standard Error	B	C
A. Outpatient	362	4532	666	–	3.30
B. Partial	326	3791	494		3.06
C. Inpatient	150	7327	1103		

Therapists' Work Setting:
Estimated Costs of Recommended Treatment and Cooperativeness

	Cooperative	Uncooperative	Standard Error	Significant Effects B	C
A. Outpatient	972	3922	708	–	–
B. Partial	926	3191	525		3.06
C. Inpatient	1162	6314	1171		

While community mental health center staff tended to give slightly higher LOF ratings overall, there was no evidence of interaction effects with any of the client patient characteristics on LOF ratings. All statistically significant interactions were observed on the length of treatment and estimated costs of recommended treatment measures. Client patient cooperativeness with treatment interacted with community center or state hospital work setting for both dependent measures (ES = 3.15 and 3.37, respectively, on the length of treatment and estimated cost measures). The employment stability interaction was observed only for the cost dependent measures (ES = 3.96).

Therapists' Work Setting:
Weighted Length of Treatment and Cooperativeness

	Cooperative	Uncooperative	Standard Error	Effect Size
State Hospital	51.81	124.82	14.54	3.15
CMHC	54.97	88.22	10.33	

Therapists' Work Setting:
Estimated Costs of Recommended Treatment and Cooperativeness

	Cooperative	Uncooperative	Standard Error	Effect Size
State Hospital	1507	7456	707	3.37
CMHC	1128	3797	1180	

Therapists' Work Setting:
Estimated Costs of Recommended Treatment and Employment Stability

	Stable	Unstable	Standard Error	Effect Size
State Hospital	631	8332	1112	3.96
CMHC	426	4499	666	

Therapists' Race. No main or interaction effects appeared to have occurred between therapists' race and race or ethnicity of client patients.

White and black therapists showed no difference in their patterns of LOF or recommended treatment judgments. The only statistically significant interaction that we found suggests that black clinical staff tend to recommend more expensive treatment for client patients with a history of unstable employment (ES = 2.91).

Therapists' Race:
Estimated Costs of Recommended Treatment and Employment Stability

	Stable	Unstable	Standard Error	Effect Size
Blacks	858	8136	1062	2.91
Whites	264	4885	730	

Discussion of Results. The study described here indicated that certain therapist characteristics should be statistically, experimentally, or administratively controlled in an evaluation research study. Whether statistical, experimental, or administrative control is used depends both on the evaluation questions raised and on the demand characteristics of the therapist variable of concern. For example, two of the therapist variables — those related to training and experience — could be controlled administratively. Lower level of functioning rated reliabilities and differences in treatment recommendations could probably be adjusted by careful training and supervision from more experienced staff. Case conferences or clinical staffing meetings could be used to this end. After a preliminary presentation of the client patient's phenomenological characteristics, staff members could be asked to write down their clinical assessment and treatment recommendations for the client patient in question. Next, the more experienced staff could be asked to present their recommendations and discuss the similarities and differences in their judgments. Assuming that the supervisees and the supervisors have agreed to use this medium as a training process with the objective of increasing agreement, then fairly high levels of interrater reliability should be able to be reached within a month or two.

For the differences found as a function of therapists' theoretical orientation, administrative procedures are probably not going to have an immediate impact. Differences in perspective usually come about after considerable amounts of clinical training, and they tend to be firmly entrenched. Thus, it is probably wise in any one evaluation research study to treat therapists' theoretical orientation as a covariate in statistical analysis of the results. One could hope that it would be possible to initiate a long-term discussion among those of differing orientations who worked in the same setting so as to attenuate the differences in their perspectives over time. The result could be that clinical staff would generate a more consistent set of therapeutic interventions. This consistency by itself may net more positive and cost-effective outcomes if the results of the research by Sachs (1983) can be extrapolated to the current discussion. Sachs found that better outcomes were obtained in client patient functioning when therapists were more consistent in their theoretical approach.

The variable of work setting (inpatient, partial hospitalization, or outpatient) is another potentially significant source of interactive variance. As with the theoretical orientation variable, training procedures could be devised to attenuate the effects of work setting interactions. However, this may not be practical or even desirable. The differences in perspective that arise from the work setting might actually foster administrative and therapeutic success of the respective settings. The fact that a client patient is functioning in the community in spite of an unstable employment history means different things when inpatient, partial hospitalization, and outpatient staff make therapeutic recommendations, although they generally agree on the client patient's overall level of functioning. Such differences in perspective align with the day-to-day experiences of staff in the work setting that they know best. Training designed to combat such differences would be superseded by daily, on-the-job learning experience.

We are not recommending in the case of either the work setting or the theoretical orientation variable that these sources of variance should be isolated without a description of interaction effects. The salient aspects of the interaction effects between each of these variables and the independent variables of interest in the evaluation study should be specified. Indeed, these interactive sources of variance may explain why a given program of services is not cost-effective overall.

While future research will undoubtedly uncover a number of variables that should be considered in a discussion of this nature, two other variables are strongly indicated by the research literature. One is related to group polarization and conformity effects in judgments made by participants (Kaplan and Schwartz, 1975). The other is related to the potential for overconfidence in judgment (Oskamp, 1965; Fischhoff, 1982).

Overconfidence in judgment is potentially a powerful negative influence that can have broad effects on the data gathered by an internal evaluation study. Oskamp (1965, p. 261) describes the overconfidence effect as follows: "(1.) Beyond some early point in the information-gathering process, predictive accuracy reaches a ceiling. (2.) Nevertheless, confidence in one's decisions continues to climb steadily as more information is obtained. (3.) Thus, toward the end of the information-gathering process, most judges are overconfident about their judgments." Kaplan and Schwartz (1975) found that, when a certain level of confidence is attained, the judge tends to stop identifying new information as new. Instead, any new data are typically seen as further support for established judgments. Thus, the judge becomes increasingly confident that the additional information recommends even greater confidence in his or her judgments, even if their predictive value is less than perfect. Indeed, Oskamp found that confidence continued to increase steadily, even when the judge's accuracy remained at 28 percent — only 8 percent above chance. Arkes (1981) reviewed the literature and the results of his own studies and concluded that, even when feedback conflicts with an established judgment pattern, it is very rare that the judgment pattern changes.

One place where the negative influence of overconfidence is potentially influential is when individual judgments are compared in order to generate a single group judgment, as when a multidisciplinary team develops a treatment plan. The research on group polarization and conformity effects is detailed in Kaplan and Schwartz (1975). For our discussion, we will simply extrapolate the implications of their findings to the issue of collecting internal evaluation data in mental health services and similar settings. (The reader is cautioned that the ensuing discussion is not directly based on empirical evidence. Rather, it is a logical extrapolation from empirical data, and the argument must be considered to be empirically weak.)

Many licensing agencies, third-party payers, and accreditation agencies require clinical input by two or more clinicians (for example, peer review, clinical supervision by a licensed practitioner, utilization and quality assurance review). From Kaplan and Schwartz, we learn that differences in clinical judgments among individual participants would be maintained during most of these interactions, even if the official group judgment was different from that of the participants. Polarization effects occur when participants begin to discuss their differences with increasing levels of intensity; that is, when they strongly disagree with one another. When such disagreement occurs, there is a tendency for one or more participants to polarize their judgments in a direction opposite to that of their opponents. While the circumstance of relative power in the organization could (and probably does) have the result that less powerful staff conform to the stand taken by more powerful individuals, the conforming party or parties do not necessarily change their views. In fact, the studies by Kaplan and Schwartz indicate that the conforming parties usually do not change their views. Moreover, and probably most important, if the arguments during the group's deliberation are sufficiently intense, it appears that the group decision will conform to the leader's polarized judgment. That is, the direction of the final group judgment is opposite that of the leader's antagonists, at a value more extreme than the leader's original judgment. Eventually, this tendency could cause group cohesiveness and willingness to cooperate to deteriorate (Argyris, 1975).

The issue that remains is whether, and how, we can rectify the judgment patterns of those who generate biased data required for internal evaluation. On the one hand, it might be statistically sufficient to enter such sources of variance into a predictive regression equation as a covariate to be partitioned out of consideration. On the other hand, one might seek to modify the clinical judges so that these differences are not evident. As noted in Chapter Four, these differences are not only organizationally devisive, but they can result in poor clinical outcomes as well.

As described earlier, simple feedback does not appear to have the desired effect (Arkes, 1981; Newman, 1983). However, recent research indicates that all is not hopeless. When judges are asked to state the potential risks of good or bad outcomes, given the known facts, before making a judgment, it appears that feedback does have a positive influence on modifying incorrect judgments (Tversky and Kahneman, 1983; and the literature reviewed in Newman, 1983).

The central theme of this chapter is that it is possible to detect the influence of therapist factors that interact with client patient characteristics to affect the dependability of clinical data. The principal variables that are related to differences in actual client patient cost outcome results have been identified. This means that, if we can identify the basic characteristics of both the therapists and the client patients, it is indeed possible to adjust the data sufficiently to perform many kinds of internal evaluation studies.

References

Argyris, C. *Intervention Theory and Method: A Behavioral Science View.* Reading, Mass.: Addison-Wesley, 1975.

Arkes, H. R. "Impediments to Accurate Clinical Judgment and Possible Ways to Minimize Their Impact." *Journal of Consulting and Clinical Psychology,* 1981, *49,* 323–330.

Fischhoff, B. "For Those Condemned to Study the Past: Heuristics and Biases in Hindsight." In D. Kahneman, P. Slovic, and A. Tversky (Eds.), *Judgment and Uncertainty: Heuristics and Biases.* New York: Cambridge University Press, 1982.

Giladi, D., and Newman, F. L. "Client Information and Its Use by Therapists Differing in Theoretical Orientations." *Psychotherapy Theory, Research, and Practice,* in press.

Heverly, M. A., Fitt, D. X., and Newman, F. L. "Constructing Case Vignettes for Clinical Judgment Research." *Evaluation and Program Planning,* in press.

Kaplan, M. F., and Schwartz, S. (Eds.) *Human Judgment and Decision Processes.* New York: Academic Press, 1975.

Myers, J. L. *Fundamentals of Experimental Design.* (3rd. ed.). Boston: Allyn & Bacon, 1979.

Newman, F. L. "Strengths, Uses, and Problems of Global Scales as an Evaluation Instrument." *Evaluation and Program Planning,* 1980, *3,* 257–268.

Newman, F. L. "Therapist's Evaluation of Psychotherapy." In M. Lambert, E. Christanson, and S. DeJulio (Eds.), *Assessment of Psychotherapy Process and Outcome.* New York: Wiley, 1983.

Newman, F. L., and Rinkus, A. J. "Level of Functioning, Clinical Judgment, and Mental Health Service Evaluation." *Evaluation and the Health Professions,* 1978, *1,* 175–194.

Newman, F. L., and Sorensen, J. E. *Integrating Clinical and Fiscal Management Systems.* Norwood, N.J.: Ablex, in press.

Oskamp, S. "Overconfidence in Case-Study Judgments." *Journal of Consulting Psychology,* 1965, *29,* 261–265.

Sachs, J. S. "Negative Factors in Brief Psychotherapy: An Empirical Analysis." *Journal of Consulting and Clinical Pscyhology,* 1983, *51,* 557–564.

Tversky, A., and Kahneman, D. "Extension Versus Intuitive Reasoning: The Conjunction Fallacy in Probability Judgment." *Psychological Review,* 1983, *90* (4), 293–315.

Frederick L. Newman is coordinator of research, evaluation, and information systems at Northwestern Institute of Psychiatry. He has conducted program evaluation studies for over fifteen years in a wide variety of settings.

Mary Ann Heverly is assistant professor at the Medical College of Pennsylvania and director of institutional research at Delaware County Community College.

Mitchell Rosen is coordinator of statistics at the University of Pennsylvania Medical School.

S. Mark Kapta is assistant professor of psychology in the Department of Psychiatry at the Medical College of Pennsylvania.

Rebecca Bedell is with the Office of Information Systems and Evaluation in the New Jersey Division of Mental Health.

Program unilization and service utilization studies form the
foundation for internal evaluation in human service organizations.

Program Utilization and Service Utilization Studies: A Key Tool for Evaluation

Gerald Landsberg

Any service organization, especially in an era of shrinking resources, needs to evaluate its services and activities. Through these evaluative activities, an organization can develop and maintain the flexibility needed to respond to an ever-changing environment. It has been suggested that, even in an ideal world, an organization needs to be self-evaluating. Self-evaluation requires an organization continuously to review its own activities and goals and to use the results to modify, if necessary, its programs, goals, and directions.

Within the agency, the essential function of evaluation is to provide data on goal achievement and program effectiveness to a primary audience consisting of administration, middle management, and governing board. This primary audience, especially the administration and the board, is frequently confronted with inquiries from important sources in the external environment, such as legislators and funding agencies. These inquiries often focus on issues of client utilization, accessibility, continuity, comprehension, outcome or effectiveness, and cost.

The building block of this information is the patterns of use or the client utilization study. The patterns of use study, whether it consists of simple inquiries or highly detailed, sophisticated investigations, is basically a descrip-

A. J. Love (Ed.). *Developing Effective Internal Evaluation.* New Directions for
Program Evaluation, no. 20. San Francisco: Jossey-Bass, December 1983.

tion. It describes who uses services and how, and it becomes evaluative when it is "related to the requirements or purposes of the organization" (Binner, 1975, p. 345). Patterns of use studies have been defined in the following terms: "[They] describe many different aggregations of CMHC structure and process and administrative outcome data. Such data can be 'patterned' from a number of viewpoints. All involve some analyses and interpolation of client characteristics, services provided, catchment area subpopulations and social areas, staff allocation, and the movement of client and staff within the center or among its activities" (Hagedorn and others, 1976, p. 123). "A client utilization reporting system provides a basis for answering such questions by systematically monitoring who comes into contact with the various systems or program units of the agency and what happens to these individuals as they maintain contact. This involves categorizing and coding characteristics of clients as well as *units* of services provided" (Elpers and Chapman, 1973, p. 15).

The information developed by patterns of use studies can be useful for a number of specific purposes, in addition to the generalized value just described: It can help to determine the extent to which high risk groups are using services. It can allow more efficient responses to client population needs. It can provide staff with significant information on their caseloads and activities. It can provide information on client movement and continuity of care issues. Most importantly, such studies provide the basis for detailed, sophisticated evaluative or research efforts on outcome, cost, cost outcome, and community impact.

Types of Patterns of Use Studies

Cross Sectional Views of Who Uses What Services. The simplest way of examining patterns of service use is to determine either at one time or over a period of time the services that are being used and the characteristics of those who use them. Comparison of this information with the demographic characteristics of the service area population and with estimates based on standardized health or mental health formulas of needs for specific types of services provides one measure of the accessibility and acceptability of a mental health program. In particular, studies of use patterns can reveal the extent to which such high-risk populations as low-income adolescents and children, the elderly, ethnic minority groups, the chronic mentally ill, and alcohol or drug abusers are using the services of a given facility. In addition, these data can pinpoint inappropriate or wasteful patterns of use. Evaluation of patterns of use can help the facility to direct its services more effectively to the community's needs.

Utilization rates, which are expressed as the number of clients served per 1,000 residents of the geographic area, are often used to describe and compare patterns of use. Rates for specific age, income, and ethnic groups are compared with known and available standards for utilization of both health and mental health services.

Studies examining patterns of use can also focus on characteristics of the interaction of staff and clients. For example, the interactions of race, age, and sex of staff and clients can affect the continuity of care associated with particular combinations of staff, and client variables can reveal barriers to treatment for certain client populations. Such data can be examined with comparable data from other agencies to clarify further the unique dimensions of staff-client interactions at the facility.

Goodman and Hoffer (1979) describe a study in which census data were used to review the characteristics of clinical service recipients. Sociodemographic data on each census tract within the clinic's catchment area were analyzed, and high-risk groups were identified on the basis of indicators from the Mental Health Demographic Profile System. Four distinct social areas were established: poor, predominantly white; poor, predominantly black; middle-class, with a sizable black subgroup; and upper-middle-class, white. Data from a computerized management information system were used to study the characteristics of a one-year cohort of clinic admissions. The characteristics examined included sex, age, ethnicity, mental status, and ideation. Admission rates per 1000 population were calculated. Data were analyzed, and the following observations were made: Blacks were admitted at a rate of four times greater than that for whites. Residents of the predominantly black social area, determined as the area of greatest risk of encountering mental illness, were admitted at greater rates than residents of the other three social areas. Blacks from middle-class areas received significantly greater rates of service than blacks from the two poorer social areas. Finally, whites of low socioeconomic status were underrepresented in admissions at the clinic. Data from this study were used to support a grant request for additional staff.

Fiester (1979) focused his study on an in-depth analysis of utilization patterns of emergency room (ER) services. Fiester asked four questions: What types of persons use ER services? What are the major referral sources? What proportion of ER contacts represent true psychiatric emergencies? What happens to persons subsequent to their ER contact? Over a two-week period, ER staff completed a contact sheet on each client encountered. In addition, clients' medical records were analyzed. The study observed a recidivism rate after two months of 34 percent. Moreover, more than half of the ER contacts were known clients, and almost one half of the contacts were nonemergency cases. This study led eventually to the creation of a separate emergency services unit.

Longitudinal Studies Tracking Clients' Use of Services. A more difficult but also a more revealing approach to the study of patterns of use tracks clients' use of services over time. Just as it is important to determine factors that promote or reduce client use of center services, it is necessary to analyze the movement of clients through a facility to ascertain the extent to which the facility's structure promotes or inhibits the appropriate and orderly use of services. One way of assessing client progress through the agency is to define a group or cohort of interest and then to monitor the movement of individuals in that group as they pass through the agency's structures. Bass and Windle (1972) used this approach to illustrate various measures of continuity of care. Key process variables that can affect the smooth transition of patients through an agency include source of referral, length of wait prior to receiving help, reception of the client, preparation of the client, and client acceptance of the treatment.

Dropout studies, which report and compare the rates at which clients withdraw from service agencies prior to a joint decision by therapist and client to terminate treatment, can be instructive in identifying internal structures that interfere with treatment. Examination of patient variables and service variables associated with dropout rates can provide clues to more effective use of agency services.

In addition to describing the past movements of clients through a facility, the tracking of clients can also provide the agency with information that can be used to project future demand for services. For instance, such variables as length of waiting list, units of service per client, and length of service per client can help to estimate future demand.

Terhune and Schultz (1979) followed two admissions cohorts through their contacts with a CMHC. A simple manual tracking system was used to collect data on clients' sex, age, diagnosis, number and dates of prior CMHC treatment, number and dates of subsequent CMHC treatment episodes, sequence of center unit in which clients received treatment, and length of inpatient treatment if any. This study produced the following observations: The emergency room was the clinical gatekeeper; 75 percent of the clients entered through this point; of these, one third were referred to other community services. Pathways through the system were usually of three types: ER service only; inpatient services only; and ER admission, followed by inpatient service, followed by discharge. Daycare was only the third filter point, after ER and inpatient. Finally, revolving-door clients — those who recycled — constituted about half the clients being served at any particular point. This study led to a series of recommendations for programmatic changes given to the CMHC administration.

Hammer (1976) examined the utilization of services by one cohort during a single year both at a CMHC inpatient unit and at the local state hospital. Hammer recorded key characteristics (age, sex, diagnosis, date, and length of stay) of patients admitted to the unit and the person's inpatient and state hospital admission data for the year. Data were collected to answer three questions: Who uses services? Are there differences in utilization by people with different characteristics? Are there any similarities among high utilizers? Data from the study showed that 65 percent of the patients were under fifty years of age, 85 percent white, 62 percent of all inpatients were readmissions and accounted for 72 percent of all admissions, and 20 percent of those admitted to the CMHC inpatient unit were also admitted to the state hospital. A closer look at the recividist population showed very different types of admission patterns that did not permit generalizations. Information from the report provided important data for programmatic considerations.

Studies Relating Utilization Data to Outcome, Cost, Cost Outcome, and Community Impact Research. As already noted, patterns of use studies are often combined with other evaluation techniques. It is common in mental health evaluation to employ utilization data with sets of outcome measures, such as community adjustment and rehospitalization rates for an aftercare program, and it is common in mental health planning to rely on utilization data that provide information about psychiatric symptomatology, behavior problems, functional skills, and nursing care requirements to project the need for inpatient psychiatric beds. Within the health arena, a range of utilization information can be combined with cost data and with patient satisfaction data to develop a hospital performance measure.

Smith and Wilensky (1979) examined a cohort of clients for key variables associated with the intake process and the outcome of therapy. The study used a specially developed instrument to collect data on clients applying for therapy. Demographic data included age, sex, income, occupation, education, and severity of disturbance. The last factor was measured by administration of the Structured and Scaled Interview to Assess Maladjustment (SSIAM). Patient pressure for service was measured by client telephone calls to therapists. Criterion measures consisted of both the calendar days (actual delay) between application and scheduled day of the first appointment and the number of therapist working days (to take into account the fewer number of days worked by part-time therapists). Failures to keep first appointments were recorded. To measure outcome, Strubb questionnaires were administered to patients and therapists after twelve sessions or earlier, if the patient terminated. Data from this study showed that the median number of days that a patient waited for service was 50. If a patient called and spoke to a professional, he waited only seventeen days. If he called and left a message, the waiting time was forty-eight days. With no call, the average wait was seventy-six days — more than two months. Higher income and education were related to keeping a first appointment. Most important, a long waiting time was associated with failing to keep the first appointment. Outcome was negatively associated with patient education, severity of disturbance, and delay in receiving treatment. Recommendations based on the study were suggested to the clinic.

Willer (1979) employed utilization study activities together with a number of outcome measures to evaluate an aftercare program. A cohort of 100 inpatients was selected to be followed through the inpatient and aftercare program. The data on patients' characteristics included reason for admission, previous hospitalization record, sex, marital status, and living and work situation. Outcome was measured by a client satisfaction questionnaire completed by the patient on discharge, goal attainment rating scales completed by staff, and follow-up interviews with patients using a community adjustment questionnaire. Data from the study showed that females were more likely to seek services. First admissions were more likely to go to other programs than to the program for aftercare. The clients who were most likely to be hospitalized were the least likely to receive aftercare. Finally, comparison of different types of aftercare and aftercare from a community-based service that offered no aftercare indicated virtually no difference between the aftercare received and the client's subsequent community adjustment. Recommendations for changes in the program were made to clinical and administrative staff.

Judging Agency Ability to Conduct Client Utilization Studies

An agency's interest and willingness to engage in client utilization studies say little about the agency's ability to undertake such activities. Mitchell (1977, p. 38) provides a checklist to judge agency capability:

1. Does your agency have a systematic way of describing clients, by classifying them into categories along relevant dimensions (for example, along the dimension of age: preschool, school-age, adolescent, adult, senior citizen; along the dimensions of sex and race)?
2. Does your agency have a systematic way of counting the number of clients contacted in terms of the particular dimensions you use to classify clients (for example, number of adolescents seen, number of drug abusers seen, and number of black women seen)?
3. Does your agency have a systematic means of counting the number of service contacts each program unit has, so that every time a client is served, this event is tabulated (for example, number of consultation visits made by consultation and education unit, number of therapy sessions in outpatient)?
4. Does your agency have a systematic means of combining information on the types of services with categories of clients so that the kinds of clients in any one service can be specified (for example, outpatient clinic sees x percent females, y percent males; inpatient has x percent schizophrenics, y percent neurotics; group therapy given involves x percent adolescents, y percent adults)?
5. Does your agency have a systematic means of combining information on the types of services with categories of clients so that the kinds of services provided to any one category of clients are specified (for example, the services provided to adolescents are x percent group therapy and y percent individual therapy; the services provided to blacks are x percent inpatient care and y percent outpatient care)?
6. Does your agency have a systematic means of keeping track of the amount and types of services provided to each client?
7. Is there a formal and continuous mechanism for disseminating the information in the previous questions to both administrative and clinical staff?

8. Does your agency have a systematic way of comparing the high-risk categories of clients identified in the needs assessment with the types of clients actually using the services of the agency (for example, are there written documents, periodic review committees, or officials designated to look for discrepancies)?

9. Does your agency have a systematic way of recording the number of referrals each agency subprogram makes and receives (for example, adolescent unit has x number of referrals, and outpatient unit makes y number of referrals)?

10. Does your agency have a systematic way of recording the number of referrals made by a service unit to a particular source or received by a service unit from a particular source (for example, adolescent unit has x number of referrals from Division of Social Services, and outpatient unit has made y number of referrals to State Hospital)?

11. Does your agency have a systematic way of following clients you refer (or who are referred to you) to see if the contact is made?

12. Does your agency have a systematic way of counting the number of referrals to a particular source or from a particular source when contact is not made (for example, x percent of referrals from Emergency Clinic do not make contact, and y percent of referrals to Division of Social Services are not made)?

If the capability exists, the agency is encouraged to pursue these activities. If the capability is not present, the agency is encouraged to develop the capability.

The Value of Patterns of Use Studies

The value of studies of patterns of use is readily evident. Such studies are easy to implement, and they enable evaluation staff to provide boards and managers with important analytic and evaluative data in a short time. The data provided by these studies often answer important internal administrative and program concerns, and they permit response to inquiries from funding and monitoring agencies. Moreover, although studies of patterns of use do not provide definitive answers about the quality and effectiveness of service, they do suggest areas for further evaluation study.

References

Bass, R. D., and Windle, C. "Continuity of Care: An Approach to Measurement." *American Journal of Psychiatry,* 1972, *129,* 196–201.

Binner, P. "Program Evaluation." In S. Feldman (Ed), *The Administration of Mental Health Services.* Springfield, Ill.: Thomas, 1975.

Elpers, J. R., and Chapman, R. L. "Management Information for Mental Health Services." *Administration in Mental Health,* Fall 1973, 12–25.

Fiester, A. "Utilization of Emergency Services at the CMHC of Palm Beach County." In G. Landsberg, W. Neigher, R. Hammer, C. Windle, and J. R. Woy, *Evaluation in Practice: A Sourcebook of Program Evaluation Studies from Mental Health Systems in the United States.* DHEW Publication No. ADM78-763. Rockville, Md.: National Institute of Mental Health, 1979.

Goodman, A., and Hoffer, A. "Mental Health Service to a Suburban Population Mixed Ethnicity and Class." In G. Landsberg, W. Neigher, R. Hammer, C. Windle, and J. R. Woy, *Evaluation in Practice: A Sourcebook of Program Evaluation Studies from Mental Health Systems in the United States.* DHEW Publication No. ADM78-763. Rockville, Md.: National Institute of Mental Health, 1979.

Hagedorn, H. J., Beck, K. J., Neubert, S. F., and Werlin, S. H. *A Working Manual of Simple Program Evaluation Techniques for Community Mental Health Centers.* Washington, D.C.: U.S. Government Printing Office, 1976.

Hammer, R. "A Comprehensive Examination of Patients Admitted to the Inpatient Unit of Maimonides CMHC." In R. Hammer, G. Landsberg, and W. Neigher, *Program Evaluation in Community Mental Health Centers: A Manual with Reports from CMHCs in NIMH Region II.* New York: Argold Press, 1976.

Hammer, R., Landsberg, G., and Neigher, W. *Program Evaluation in Community Mental Health Centers: A Manual with Reports from CMHCs in NIMH Region II.* New York: Argold Press, 1976.

Mitchell, R. "The Dimensions of an Evaluation System for Community Mental Health Centers." In R. Coursey, G. Specter, S. Marrell, and B. Hunt, *Program Evaluation for Mental Health: Methods, Strategies, and Participants.* New York: Grune & Stratton, 1977.

Smith, G., and Wilensky, H. "Intake Priorities and Effects in Early Psychotherapy." In G. Landsberg, W. Neigher, R. Hammer, C. Windle, and J. R. Woy, *Evaluation in Practice: A Sourcebook of Program Evaluation Studies from Mental Health Systems in the United States.* DHEW Publication No. ADM78-763. Rockville, Md.: National Institute of Mental Health, 1979.

Terhune, K., and Schultz, M. "Client Processing at the Niagara Falls CMHC: A Systems Analysis." In G. Landsberg, W. Neigher, R. Hammer, C. Windle, and J. R. Woy, *Evaluation in Practice: A Sourcebook of Program Evaluation Studies from Mental Health Systems in the United States.* DHEW Publication No. ADM78-763. Rockville, Md.: National Institute of Mental Health, 1979.

Willer, B. "Evaluation of a Hospital-Based Aftercare Program." In G. Landsberg, W. Neigher, R. Hammer, C. Windle, and J. R. Woy, *Evaluation in Practice: A Sourcebook of Program Evaluation Studies from Mental Health Systems in the United States.* DHEW Publication No. ADM78-763. Rockville, Md.: National Institute of Mental Health, 1979.

Gerald Landsberg has been director of Ulster County Community Mental Health Services for the past four years. During the previous ten years, he was director of program evaluation for Maimonides Medical Centre, the Community Mental Health Center. Also, he served for five years as the Chairman of the Research and Evaluation Council of the Community Mental Health Center.

*Can the same person be responsible for evaluation and
implementation? What are the limits of self-evaluation?*

Should Evaluation
Become Implementation?

*Angela Browne
Aaron Wildavsky*

How do we know whether a program embodying a policy has been well or
poorly implemented? By observing the difference between intended and actual
consequences, that is, by evaluation. Indeed, by expanding evaluation beyond
the mere measure of outcomes to cover the causes of the consequences observed,
we can use such knowledge to alter programs or their mode of implementation. Whether evaluation is used to check progress or to change direction, it
involves the analysis of implementation.

Implementation involves continuous adjustments between policies and
their consequences. The evaluation component of the policy process is an
activity designed to narrow the gap between what happens and what ought to
happen. Implementation creates opportunites for policy evaluation; in itself,
implementation is a test of policy in action. In evaluating implementation,
information can be generated that can narrow the gaps between intended and
actual consequences. Thus, implementation and evaluation are the opposite
sides of the same coin, implementation providing the experience that evalua-

This chapter is a revision and condensation of material to appear in the third
edition of Jeffrey Pressman and Aaron Wildavsky, *Implementation* (University of California Press, spring 1984).

A. J. Love (Ed.). *Developing Effective Internal Evaluation.* New Directions for
Program Evaluation, no. 20. San Francisco: Jossey-Bass, December 1983.

tion interrogates and evaluation providing the intelligence to make sense of what is happening.

In the world of action, implementation and evaluation are often carried on by the same people — public officials. They act and observe, observe and act, combining program execution with intelligence about consequences so as to reinforce or alter behavior. Doing well or doing badly, hardly conscious of the analytic distinctions involved, participants in the policy process act simultaneously as evaluators of the programs that they implement and as implementers of the programs that they evaluate.

It has to be so. Even where formal attempts are made to separate policy from administration, whether it be between the legislative and executive branches or between policy analysis and program administration units, the vast bulk of activity is carried on in the field. Outside forces are overwhelmed in numbers and expertise by men and women inside the organization. Given the size and scope of contemporary government, most intelligence about events and their consequences has to come from close to the ground.

It is well to understand that formal evaluators, being once (if they work in agencies), or twice (if they are part of overhead units, like the Office of Management and Budget or the General Accounting Office), or three times (if they are outside government) removed from the scene of action, cannot substitute for or replace public officials themselves. Such replacement would be undemocratic; it would also be cognitively unfeasible. If mankind sees through a glass darkly, what shall we say of those who see almost entirely through other people's prisms?

Selling evaluation goes with the job. Making it more attractive to different constituencies or organization units is part of being persuasive. So is writing well. Why shouldn't evaluators go whole hog by superintending the implementation of program evaluation?

In the world of theory, distinctions can be made that are, of necessity, blurred in practice. There, if we have the wit, we can control concepts instead of being dominated by them. In the world of abstractions, evaluation can be distinguished from implementation. The question is whether it is intellectually satisfying and practically useful to do so.

By blurring the distinction, so that evaluation becomes preoccupied with utilization, and implementation takes up monitoring and causality, the two subfields merge into a single-seamed concern with policy analysis. Designing policies to be evaluable and implementable, discovering new alternatives, bringing in new values and constituencies, recommending choices among possible programs, setting up coalitions to support preferred programs, and more, much more, ranging from establishing political feasibility to devising organizational incentives — all these things become parts of the task of evaluators who are also designers, organizers, strategists, and politicians. A sense of power and responsibility is gained. A sense of differentiation, appropriateness, and hence professionalization is lost.

Evaluators have tried to become implementers. This is both good and bad. It is good because evaluation becomes more relevant. It is bad because evaluation becomes less knowledgeable. Being useful is one thing; acting as if one were the implementer is another. Monitoring the consequences of programs, which up until now has been the core of evaluation, requires a certain willingness to follow where the evidence leads. The evaluator must always be prepared to discover that the consequences (or the objectives) under consideration are not the only ones. Such breadth of vision is sacrificed when evaluation becomes indistinguishable from implementation or intelligence becomes indistinguishable from action. Where an absence of interest in utilization is stultifying, an exclusive concern with immediate use results in the subordination of intelligence to action. In the end, if this path is taken, only self-serving evaluations will be made. Why care, then, whether evaluations whose conclusions have been prejudged are used or not? If we do care, it is only because we want to see as little of this sort of "evaluation" as possible. Demarcating the domain of implementation helps to maintain the distinctiveness and integrity of evaluation.

Evaluators, who cannot take objectives as fixed, for otherwise there would be no need for them, need to remain flexible. Thus, they are wise not to fix prematurely on immediate utilization in the flux of life. They are wise also to infuse their own evaluations with multiple possibilities, some of which may be more important in the future than they are now. The conceptual distinction between evaluation and implementation is important to maintain, however much the two overlap in practice, because it protects against the absorption of analysis into action, to the detriment of both.

Angela Browne is affiliated with the School of Social Welfare at the University of California at Berkeley.

Aaron Wildavsky is professor of political science and public policy at the University of California at Berkeley. For over a decade, he has been interested in the self-evaluation of public programs.

Index

105

Statement of Ownership , Management, and Circulation
(Required by 39 U.S.C. 3685)

1. Title of Publication: New Directions for Program Evaluation. A. Publication number: 449-050. 2. Date of filing: 9/30/83. 3. Frequency of issue: quarterly. A. Number of issues published annually: four. B. Annual subscription price: $35 institutions; $21 individuals. 4. Location of known office of publication: 433 California Street, San Francisco (San Francisco County), California 94104. 5. Location of the headquarters or general business offices of the publishers: 433 California Street, San Francisco (San Francisco County), California 94104. 6. Names and addresses of publisher, editor, and managing editor: publisher—Jossey-Bass Inc., Publishers, 433 California Street, San Francisco, California 94104; editor—Ernest House, CIRCE-270, Univ. of Illinois, Champaign, IL 61820; managing editor—William E. Henry, 433 California Street, San Francisco, California 94104. 7. Owner: Jossey-Bass Inc., Publishers, 433 California Street, San Francisco, California 94104. 8. Known bondholders, mortgages, and other security holders owning or holding 1 percent or more of total amount of bonds, mortgages, or other securities: same as No. 7. 10. Extent and nature of circulation: (Note: first number indicates average number of copies of each issue during the preceding 12 months; the second number indicates the actual number of copies published nearest to filing date.) A. Total number of copies printed (net press run): 3674, 3645. B. Paid circulation, 1) Sales through dealers and carriers, street vendors, and counter sales: 85, 40. 2) Mail subscriptions: 1469, 1170. C. Total paid circulation: 1554, 1210. D. Free distribution by mail, carrier, or other means (samples, complimentary, and other free copies): 125, 125. E. Total distribution (sum of C and D): 1679, 1335. F. Copies not distributed, 1) Office use, left over, unaccounted, spoiled after printing: 1995, 2310. 2) Returns from news agents: 0, 0. G. Total (sum of E, F1, and 2—should equal net press run shown in A): 3674, 3645. I certify that the statements made by me above are correct and complete.

JOHN R. WARD
Vice-President